T0332289

Water Poverty

The Next "Oil" Crisis

Shirley J. Hansen, Ph.D.

River Publishers

Routledge
Taylor & Francis Group
LONDON AND NEW YORK

Published 2020 by River Publishers

River Publishers

Alsbjergvej 10, 9260 Gistrup, Denmark

www.riverpublishers.com

Distributed exclusively by Routledge

4 Park Square, Milton Park, Abingdon, Oxon OX14 4RN

605 Third Avenue, New York, NY 10017, USA

First issued in paperback 2023

Library of Congress Cataloging-in-Publication Data

Names: Hansen, Shirley J., 1928- author.

Title: Water poverty : the next "oil" crisis / Shirley J. Hansen, Ph.D.

Description: Lilburn, GA : Fairmont Press, Inc., [2016] | Includes
bibliographical references and index.

Identifiers: LCCN 2016020838 (print) | LCCN 2016027318 (ebook) | ISBN
0881737593 (alk. paper) | ISBN 9788770223355 (electronic) | ISBN
9781498796439 (Taylor & Francis distribution : alk. paper) | ISBN
9788770223355 (Electronic)

Subjects: LCSH: Water supply. | Water consumption. | Water security. | Energy
consumption. | Energy development.

Classification: LCC HD1691 .H33537 2016 (print) | LCC HD1691 (ebook) | DDC
333.91--dc23

LC record available at https://lccn.loc.gov/2016020838

Water poverty / Shirley J. Hansen.

First published by Fairmont Press in 2017.

©2017 River Publishers. All rights reserved. No part of this publication may
be reproduced, stored in a retrieval systems, or transmitted in any form or
by any means, mechanical, photocopying, recording or otherwise, without
prior written permission of the publishers.

Routledge is an imprint of the Taylor & Francis Group, an informa business

Publisher's Note
The publisher has gone to great lengths to ensure the quality of this reprint but
points out that some imperfections in the original copies may be apparent.

ISBN 13: 978-87-7022-937-1 (pbk)

ISBN 10: 0-88173-759-3 (The Fairmont Press, Inc.)

ISBN 13: 978-1-4987-9643-9 (hbk)

ISBN 13: 978-8-7702-2335-5 (online)

ISBN 13: 978-1-00-315202-6 (ebook master)

While every effort is made to provide dependable information, the publisher, authors, and
editors cannot be held responsible for any errors or omissions.

The views expressed herein do not necessarily reflect those of the publisher.

Table of Contents

Introduction

When a small child is forced to drink contaminated water in order to stay alive, we all cringe. That is poverty of the worst kind. A water crisis on our immediate horizon is destined to hurt, even kill, millions of children, and the window of opportunity to do something about it is rapidly closing.

Water is so vital to our personal well being, industrial growth, and the environment, that a shortage could tear us apart morally and physically. Through the "energy shortage" years, we took comfort in finding alternative energy and some Yankee ingenuity to fill in the gaps. But there is no "alternative" water and we see neither Yankee, nor Southern, ingenuity to put the H_2 and the O together to solve the coming water crisis.

There is, however, a glimmer of hope that could turn into rays of sunshine. Water is a commodity, and we have just come through some painful times dealing with the shortage of another commodity, energy. We have faced some staggering problems and found some surprising solutions along the way.

For those who lived through the "energy crisis," the first chapters in this book offer a brief trip down memory lane. For those who must address the coming water crisis, you are offered a guided tour with some rare glimpses into some striking similarities we can draw from our energy experiences. More than once we stubbed our toes and had some truly painful experiences. In retrospect, those experiences could help us cope with the water poverty that is fast approaching. Even better there is probably no group better equipped to deal with the challenges ahead than energy engineers. After all, we lived it.

This book is designed to examine our energy history and glean some helpful hints to bolster our efforts to address water poverty. Starting with the 1970s, when we were caught by surprise in what we called an "embargo," there are some incredible lessons learned from our missteps and bureaucratic fumbles. The opportunity exists to use what we have learned and lead our nation and even the world through some potentially dark days ahead into a truly water sustainable future.

In preparing this book, I am once again deeply indebted to James Christian Hansen, who was a great sounding board, edited most of the chapters, and even provided the title, and to our son James Douglas Hansen, an exceptional writer/editor who proudly follows in his father's footsteps. Sincere thanks to Joy Maugans, who takes a rough manuscript and turns it into what you have in your hands. I also wish to express my sincere appreciation to James W. Brown, P.E. and Dr. Jennifer L. Bon for their review and comments. It is a bold endeavor to address such a momentous threat to our society, our very way of life. It is my fondest, even desperate, wish that the words here will sow the seeds for the planning we critically need to keep our world from literally draining away from under our feet.

Good luck, my friends.

Shirley J. Hansen

Chapter 1

The Birth of an Energy Expert

When a crisis hits, we look for help where we can get it. When the oil crisis hit, no one was prepared.

While many in the energy field view the "oil embargo" of 1973 as heralding in the new age of energy consciousness, for me it truly began when I was asked, "How do we know engineers can predict energy savings?"

Those asking were members of Congress and their staffs. The year was 1975, and the answer to the question at the time was, "We don't."

TURNING THE CORNER

Given the benefit of 20-20 hindsight, we can sometimes pick the moment we turned THE corner—and our world suddenly spun off in another direction. We seldom know at the time that we have hit an event that will change our lives. If you think back to some small event that turned out to have a huge impact in your life, you'll probably agree that those corners should come with big flashing neon signs.

We have all had some fascinating corners that have shaped our lives, but they seldom come out of the blue and crack you upside the head like this one hit me.

It all started on a quiet spring morning in the offices of the American Association of School Administrators (AASA). I was about to tackle some exciting legislative issue like school lunch when I was told I was needed in the second-floor conference room—a room reserved for big, important meetings.

I had recently donned the mantel of "legislative specialist" (at the time that was education speak for lobbyist) for AASA,

1

the nation's school superintendents' group. I entered the room with some trepidation—the new kid on the block. My boss, a great guy named Jim Kirkpatrick, proceeded to introduce me to a visionary named Ed Stephan. It seemed the meeting had been called to address the nation's school superintendents' number-one concern: escalating energy costs. As the conference proceeded, I discovered I had been anointed as the "go to" girl to resolve the public schools' energy and associated budget problems.

I didn't know a thing about *energy*.

My marching orders were to go up on The Hill (Congress) and see if we could do something to help our schools manage their escalating energy costs. Suddenly, everyone in the conference room was looking at me like I could pull a kilowatt out of a hat.

Now *that* was a corner! I suspect 'til this day you can see the skid marks on the conference room floor where I made an abrupt turn. What a change in direction—from school lunch and copyright laws to ENERGY. As I looked around that conference room in something of a daze, I suddenly realized it had been collectively decided that I was the closest thing we had to an energy efficiency expert. At least, it became quickly evident that I'd better become one—and fast.

Nobody—but nobody—knew a thing about "energy," so I had lots of company. Until 1973, most of us thought of energy in conjunction with words like *kinetic*. Or, if we were truly sophisticated, we might have known a little bit about horsepower. Most of us didn't know a Btu from a hole in the ground, or why we suddenly had to pay higher light bills.

Even worse, school people had no idea where they could get the money to pay those utility bills.

Unlike every other sector of the economy, the public schools could not raise prices to cover their climbing electricity costs. In the 1970s, school boards typically set the budget a year ahead, and state appropriations, which represent a huge portion of school revenues, were (and are) often set at least a year before that (in some states two years earlier). A quick look around reveals that the schools still face the same dilemma today.

In the 1970s, the problem was compounded by the fact that 75 percent of the U.S. school buildings in America had been built to house the "baby boomers." Freely translated, that means low-cost housing. Or, to put it more crudely, "cheap." Constructed out of little more than *papier mâché* and toothpicks, our schools were glass box energy sieves. In fairness to the architects and engineers as well as the school boards that hired them, the challenge was to create the square footage necessary to house and educate all those post World War II urchins, which were suddenly underfoot. And energy was cheap. Few in the 50s and 60s worried about the cost of heating or cooling that square footage.

It seemed like overnight someone closed the valve on the oil pipelines. Long lines formed at the gas pump; 26¢ per gallon at the local gas station was suddenly history. School superintendents were fighting rigid budgets. To pay the utility bills and fuel those school buses, they had little choice but to cut education programs. School people enter the profession because they want to help kids learn. Cutting into programs that might help a young boy or girl learn some critical skill is an anathema to educators. So it was not too surprising that when the school superintendents were polled by AASA in 1975, they named climbing energy costs as their number-one concern. The problem was what to do about it.

ENERGY EMBRYONICS

As the folks in that AASA conference room fervently looked about for someone with energy expertise, it became increasingly evident that everyone was looking at me. Then, in a sudden moment of realization, like the proverbial light bulb going off, I realized why they were looking my way. In full panic mode, I took a surreptitious look around the table. If there was ever an energy embryo, I was it. Obviously I had been appointed when I wasn't looking. The tenor of the meeting, however, made it abundantly clear that it was expected that I would do something about the problem.

And that is how energy experts are born.

Now, if we fast forward, the question is: How can we give birth to some water experts?

Chapter 2

The Energy Efficiency Evolution

Now, for that trip down memory lane.

The year was 1973. Elvis Presley had just made the first worldwide telecast by an entertainer. Other world-shaking events included the sale of the New York Yankees by CBS for $10 million and the winning of the Preakness by Secretariat. In many respects, it was relatively quiet and serene.

Into this peaceful scenario dropped the Oil Embargo. In a country with a mad love affair with the automobile, pain at the pump was a particularly ugly surprise. The front page of the newspapers carried pictures of cars lined up at the gas pumps like the picture below.

Headlines declared energy prices were climbing towards a dollar. Those, who were teenagers at the time, can vividly remember the sudden shock of not being able to drive just anywhere. Suddenly, there was a question as to whether the car had enough gas to go to the movies. What on earth could one do on the big date if you couldn't even cruise main street?

The panic reached every corner of our economy. The knee-jerk reactions to sky-rocketing energy prices were huge.

WHEN PANIC REIGNED

We were totally unprepared for a sudden energy shortage. President Nixon ordered the lights that flooded federal buildings to be turned off. By December, the choice was no federal Christmas tree, or one with no lights. Options were bleak. Life was grim.

The schools looked like they were braced for a perpetual hurricane. Plywood covered most of the windows. I remember driving through Lincoln, Nebraska, with the assistant superintendent of schools, who pointed to the covered windows and the boarded-up air intakes and questioned how the district could possibly pay its utility bills.

School buildings became a symbol of a world gone crazy.

The logic was obvious. Cold drafts came from around some poorly installed windows, so we began to question the amount of air that was deliberately brought in through the air intakes. Warming the air was costly. The alternative was to block the drafty windows and cover the air intakes. The utility-bill sticker shock pushed daylighting and "fresh air" concerns totally off the table.

These aberrations fed the later stories about our indoor air quality (IAQ) problems being rooted in the energy *conservation* efforts of the 1970s. As discussed in Chapter 5, there was some merit to the concern, but painting energy efficiency as the IAQ culprit was vastly exaggerated. It seems one problem breeds another. Looking ahead, it raises the specter of what associated problems will be spawned from our water crisis. It does not take much imagination to envision someone blaming a sudden flu epidemic on the *reused* water.

As the energy shock rippled across our economy, public schools made a great place to illustrate what was happening. First, school buildings represented (and still do) a microcosm of the world. From restaurants, to sports arenas, to auto shops, to auditoriums, it's all there. Second, public schools were in every community—buildings everyone was familiar with. Third, they offered a dramatic example of what happens when we are faced with a sudden dilemma and do not have the answers or the money to address the problem. And most telling, they involved little children who were required by law to be there.

School administrators were not only hit with trying to keep the lights on and keep the classrooms at a decent temperature, but all those big yellow buses needed fuel, too. Some accommodations were made in bus routes. Kids were expected to walk farther to the bus stop. Routes were consolidated. The mad scramble was to find a miracle fuel additive, which would get more miles per gallon.

By the late 1970s, President Carter had taken a page from President Roosevelt's play book and had his own version of the "fireside chats," complete with sweater. Congress passed the Emergency Energy Conservation Act, which established school building temperature restrictions. A whole generation of students can recall what it was like to wear coats in school all day. There are a few white-haired administrators, who can share stories about mounting teacher "cold classroom" frustrations. Ever try to teach a kindergartner to write while he is wearing mittens?

The "energy crisis" was pernicious and pervasive. The drive was to cut consumption. Keeping the lid on escalating utility bills was paramount. Figuring out how to use energy more efficiently would come later.

These early reactions to the energy shortage are a harbinger of things to come when we suddenly run short of something as essential to life as water. To this day, we pay for much of the confusion that reigned back then. Consider:

- It still has not penetrated that our fossil fuels are treasured and finite resources;

- "Energy Conservation" is frequently perceived as equating to deprivation;

- Energy managers are still regarded more as techies than managers. (They should be at the management table caring for one of our most valuable resources and a vital production element.)

- The energy shortage made oil concerns a critical part of our foreign relations policies;

- We moved from a vague notion of Arabs as nomads who wandered the desert with dish towels on their heads to a keen awareness of our "vital interests" in the Middle East;

- President Nixon's negotiations with the Saudis established the petrodollar, which is credited with leading to the American dollar becoming the world's reserve currency; and

- Our resistance to the feeling of dependence on others for the life blood of our economy still permeates our energy thinking today, and contributes to the "anti-fossil fuel" sentiments that fuel the environmental movement, including our rather illogical reliance on alternative fuels when energy efficiency in many circumstances is a much more economical and practical approach.

It's amazing how the ripples of panic from an unprepared nation are still reflected in our actions today. And perhaps more disturbing is the fact that, despite all the rhetoric, few nations (including the US) have a cohesive or coherent energy policy. Yet, these ripple effects are apt to have an even greater impact as we move more decisively into our sustainability efforts, or we feel the full impact of the water shortage.

Today, the US has abundant natural gas and is in a position to export our energy. Problem solved! Right?

The concept prevails that history sets the context for the present.

Just in case we need reminding, the phrase, "What is past is prologue" is inscribed in stone at the National Archives building in Washington, D.C.

This thought-provoking phrase reminds us that we have the opportunity to build on what has gone before. Previous experiences lay the foundation, maybe even dictate, what is to come. Tragically, the past also carries components of inevitability and vulnerability *if we are aware enough to spot them.*

As observed by Santayana, "Those who cannot remember the past are condemned to repeat it."

It's eerily and hauntingly familiar. It's today's equivalent of the boarding up school windows and blocking air intakes. Once more a vital resource is threatened and few, especially our public agencies, seem to know what to do. They are grasping at straws.

As we reflect on the struggles and uncertainties that surrounded us, it does not bode well for our ability to address the looming water crisis. A closer look at our response to the energy crisis will give us some critical guidance in the water problems that are lapping at our shores.

Energy problems that are hauntingly familiar keep coming back to plague us. Happily, we did some things right. Good or bad, there is still much we can gain from a brief look at yesteryear. This chapter, therefore, begins with the good ol' days when gas was 26¢ per gallon and we really could function through a whole day without a phone in our pocket.

There are those who go to great lengths to protect the past, defend past actions, and point to the "good ol' days." Some can't resist noting the way "we have always done it." Such an allegiance to the past may find us hidebound, bogged down in tradition, or relying on the false security of the familiar. Some-

Monument at the National Archives, Washington, DC

times our dedication to the past gets in the way of freeing us to try the obvious. It is dangerously close to the denial that prevails today regarding our water needs.

When do we stop watering the golf course with drinking water?

BACK THEN IS RIGHT NOW

Looking back from the second decade of the 21st century, it's hard to believe our thinking was totally focused on energy "conservation." Once the mass hysteria subsided and the smoke cleared, we realized that *conservation*—cutting back—was not, *and is not*, always the best answer. Gradually, we became aware that a wise approach to energy use was to consume the energy that must be used as efficiently as possible. During the transition years, the terms "energy conservation" and "energy efficiency" were frequently used interchangeably. Today, energy efficiency should be the preferred approach and the dominant concept. Only when we truly mean to cut back do the knowledgeable people in the industry use the term energy *conservation*.

As discussed in Chapter 5, when we practice energy efficiency; i.e., using the energy we must use as efficiently as possible, the IAQ fears (as well as deprivation fears) are clearly mitigated. I wish I had a nickel for every time a member of the school board asked me if letting an energy service company reduce energy waste meant the children would be freezing in the dark.

The uninformed and the misinformed provide fertile grounds for panic. The rush to judgment sets the stage for poor planning. As we brace for the water crisis, the deluge of problems that will confront us due to the uninformed and the misinformed should be central to our information and communications planning. We can start by recognizing the difference between potable water, grey water and black water and making the distinction part of our daily lives.

At one point in my farming days, I was subjected to the wonderful experience of getting fryers ready for the freezer. I can assure you that chickens really do run around after their heads are cut off.

In the mid- to late 1970s, we truly resembled those chickens. The panic was real, the confusion rampant, the knowledge limited.

There were beacons of light and moments of inspiration. One of those came from a visionary named Ed Stephan, who encouraged an engineer named Goody Taylor to survey some schools in Arlington, Virginia, to identify ways to reduce energy consumption. The year was 1962. Ed and Goody were way ahead of their time. The prevailing sentiment, however, was why on earth should we worry about using less of something that was so cheap? (If we thought about it at all!) But in the mid-1970s, these old surveys were dusted off and examined for money-saving opportunities.

ENERGY AUDIT EMBRYONICS

With members of Congress asking if engineers could predict energy savings, those early surveys became exceedingly valuable. They were far from perfect, but they offered a model for identifying "energy conservation opportunities." With the instigation of Ed Stephan, the foresight of one Jim Kirkpatrick and a lot of labor in the vineyards by yours truly, the American Association of School Administrators' Saving Schoolhouse Energy (SSE) project was born.

Those early surveys helped convince the doubters that we could have *planned* energy management programs. But the first step was to check the buildings for energy conservation opportunities. Once the opportunities were identified, then manufacturers such as GE, Livonia and Dow Chemical supplied lamps, fixtures, and insulation. Once installed, the energy saving benefits were assessed.

I have absolutely no idea where the term "audit" came from, and can't find anyone who knows for certain. The talk back then rested on conducting surveys that represented an accounting of what existed. It was definitely a "techie" thing, and the money overtones often associated with auditing were woven in as we went along.

But the *audit* found a home in NECPA (the National Energy Conservation Policy Act). Its "prenatal care" grew out of the AASA project. The plan was to prove that engineers could predict energy savings and, at the same time, demonstrate the value of the energy efficiency and the survey itself by "auditing" ten schools around the country. Not too subtly, those schools were placed in key congressional locations. One was in Hindeman, Kentucky. The Honorable Carl Perkins, Chairman of the House Education and Labor Committee, just happened to represent the eastern part of Kentucky where Hindeman is located. I had asked the people at the Kentucky state office of education for help in identifying a school in eastern Kentucky. When they came back to me with one in Hindeman, I was worried that this was the actual home town of Chairman Perkins and the selection might be just a bit too obvious. So I called Mr. Jack Jennings, who was Mr. Perkins' administrative assistant, to ask if the selection of Hindeman was too blatant. Jack responded very directly, "There is nothing too blatant in Kentucky."

Sometime later, I had the opportunity to visit Hindeman. On the way there, we passed the Linden Baines Johnson Elementary School. I was told it had been built with "Title One" money. Being the typical font of all wisdom from DC, I explained to the locals that Title One was for comprehensive education, not school construction. It was then explained to me that construction funds could be made available under Title One if the school were named LBJ. Nothing was too blatant in Kentucky.

As an aside, it should also be noted that this process offered me an opportunity to learn a lot more than the nuts and bolts about "energy." Hindeman was a perfect example of reminding me that, despite what the folks inside the DC beltway think, they do not have all the answers as to what is going on around the country.

Sometimes the school site selection process had a little help from the other end. The ranking Democrat on the House Appropriations Committee, Congressman Obey, got wind of the project (or the school superintendent of Stevens Point, Wisconsin, did and whispered in the congressman's ear) and a little well placed persuasion added the Plover Whiting school in Stevens Point to the list.

Another selected school was in Warwick, Rhode Island. Senator Claiborne Pell of Rhode Island was the chair of the Senate Subcommittee on Education.

Sometimes those chickens, even those with their heads cut off, come home to roost. Senator Pell recognized the merits of helping the schools manage their energy costs and asked one of his committee staff, Cary Peck (son of Gregory Peck), to work with AASA to draft some legislation. The term "energy" did not appear until the seventh page in the drafted legislative language, so the Subcommittee on Education could retain jurisdiction over the bill and hold hearings on it. The legislation actually got ahead of the project and, as is wont to happen in Washington, D.C., the law was passed before we ever got around to proving that engineers could predict energy savings. At moments like that, a good lobbyist (or legislative specialist as I was called then) just smiles … a lot.

A provision in NECPA exemplifies how uncertain we were about the audit procedures, and in fact, where the whole process was going. As part of our SSE project, we had a school in Lubbock, Texas. The audit cost was $8,000, and at the time, only $1,000 in viable energy conservation opportunities could be found. Part of the problem was the still relatively low price of energy, but most of it was that we didn't yet know what to look for and how to calculate the savings. This uncertainty and the desire to keep audit costs confined to the buildings that really needed them prompted the idea of having a Preliminary Energy Audits (PEAs) program. The primary purpose of the preliminary survey was to determine if a full-scale audit was justified.

A key part of NECPA was the Schools and Hospitals Energy Grants Program (later referred to by U.S. DOE as the Institutional Conservation Program, or ICP). Parts of the grant program provided funds for the PEA and the TA (Traditional Energy Audit), to establish state energy offices, and to train auditors.

Misinformation was rampant and the press added to the confusion. Unsure what to call the grants program, the press frequently referred to it as money for insulation projects. As every good energy engineer knows today, insulation is seldom the most cost-effective measure to install. It was part of the reason for the predicted poor payback at Brown Elementary in Lubbock.

As discussed in greater detail in Chapter 9, much political capital is made by attacking lobbyists and the nefarious work they do. If Congress, however, is going to make decisions which will affect every one of us, it surely helps if the members have some idea of what they are doing. A good lobbyist makes sure those members of Congress are making INFORMED decisions.

Amid the energy confusion in the 1970s, there was considerable uncertainty as to what should be in the Schools and Hospitals Energy Grants portion of NECPA. In an opportunity that makes lobbyists salivate, I was asked to write much of the legislative language in that particular section. Some of the golden words that would shape a whole new industry were drafted at my kitchen table. That labor of love included a definition of what should constitute an audit.

Later, as head of the task force promulgating the regulations issued in the 1981 Federal Register for the Schools and Hospitals portion of NECPA, I also had a hand in detailing exactly what an early audit should look like. Much, much later the audit was refined and we dubbed it an "Investment Grade Energy Audit." As we worked on it as co-authors of the book, *Investment Grade Energy Audits,* Jim Brown and I were amazed to realize how well the original audit had stood the test of time. Portions of it are still in use today. For a "cut and paste" effort, manufactured in a haze of uncertainty, it is amazing how much a group of outstanding engineers from across the nation were able to accomplish.

In times of crisis, *there is great value of getting knowledgeable people together at the right time to develop carefully thought out plans.* As the planning goes forward to meet our water concerns, this experience in addressing our energy needs could serve us well and very likely save much second guessing and retooling.

GETTING THE WORD OUT

Legislative language and regulations are just words on paper until someone translates them into action. NECPA provided a foundation for much of the energy development, particularly in the public sector. But the true evolution would have been stunted without the work and advocacy of one group, the Association of Energy Engineers (AEE), and more particularly the actions of one man, AEE's Executive Director, Al Thumann.

AEE'S Critical Role

It's nearly impossible to overstate the uncertainty and confusion that surrounded our early energy efforts. The need to bring some stability to the emerging profession was critical. Many professionals with good intentions did not know what to do. Too often those seeking services did not know how to judge the value of the supplier. Still others seeing a lucrative opportunity, learned the jargon and then sold the unsuspecting a bill of goods.

Into this quandary of uncertainty, the Association of Energy Engineers brought much-needed guidance. The seminars and conferences improved the caliber of services professionals offered. The certification process set critically needed standards and also provided clients a means of selecting qualified assistance.

These actions by AEE gave structure to a new industry and advanced the acceptance of energy efficiency as a response to our nation's energy needs. As we explore the uncharted waters of the emerging crisis, *the leadership of AEE, or a similar organization, could make a big difference in the orderliness of the process.*

During the late 1980s, after the price of energy fell, the energy efficiency industry suffered. As it languished, efforts to use energy more efficiently would have come to a standstill if we hadn't had the leadership that Al Thumann gave the industry, and the support he got from its education arm, The Fairmont Press. During this time and in the subsequent years, the seminars, the conferences, the certification programs, and the books all developed under AEE's leadership contributed to the high standards and quality that the industry enjoys today in the US and around the world.

By 2000, the Schools and Hospitals Energy Grants Program was credited with saving our public institutions—*and the taxpayer*—more than $15 billion. That's a huge carbon footprint. And it would not have happened without the education and guidance of AEE and the quality of service its members have provided over the years. Instead of those dollars going up the smokestack, they have gone to educate kids and provide better patient care. In the process, it has preserved critical fossil fuel reserves and reduced potential environmental damage. It is a great accomplishment in which AEE and every one of its members should take pride.

As water uncertainty mounts in this second decade of the 21st century, recalling what AEE did back then to support the energy industry could offer a valuable template for addressing our water shortage. Replicating some of the association's actions could spare us some false starts or keep us from imposing heavy regulations where a simple solution might suffice.

Other associations, such as the Alliance to Save Energy, have provided valuable leadership and service. Thanks to such groups, energy efficiency has become standard nomenclature, and standards have been established in construction, appliances, vehicles, etc. In addition, the growing use of alternative energy sources and the more conscious use of our limited resources continue to improve. We have come a long way since the mid-1970s, but we still have quite a way to go.

The beat goes on. We have only to look at the most recent data from US DOE, such as Figure 2-1 on the following page, to see how our analysis has improved. We now have a good read on who is using what. It helps us focus on what we need to address next.

Speaking of what to do next, DOE has just issued a new standard which the agency estimates will save US businesses $167 billion on their utility bills and reduce carbon pollution by 885 million metric tons when implemented. These new commercial air conditioning and furnace standards will take effect in two phases. The first phase will begin in 2018 and will deliver a 13% efficiency improvement. Five years later, an additional 15% increase in efficiency is required for new commercial units.

What makes the rule the biggest ever, according to the Energy Department, is the total amount of energy that it saves over the lifetime of the standard, which would be 15 quads—short for a quadrillion (a one with 15 zeros after it) British thermal units, or Btus.

So if we pause to assess how we did back then, we must conclude that we did remarkably well, but an awful lot of it was just fool luck. Then we stop to see where we are today and realize we are effectively using the data we collect and planning better. Now the question is: What have we learned? When faced with another shortage of an even more critical resource, such as water, will we reveal a better, more methodical approach to the problem?

PAYING THE PRICE

After the infamous "Oil Embargo" of 1973, when supplies were limited, the frantic search for ways to conserve energy dominated facility management. Amid the boarded-up windows, the dimmed lights, and orders from the federal gov-

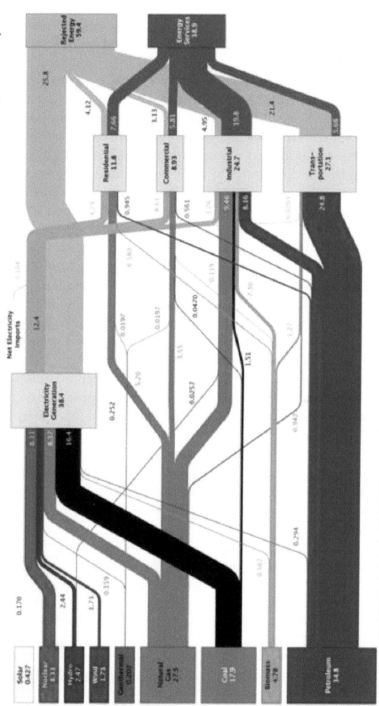

Figure 2-1

ernment to turn down thermostats came a desperate need for someone to tell us what to do to save energy. Out of this mess, came the *energy manager*.

We wanted, and needed, a "techie," and we got one.

As a result of EECA (the Emergency Energy Conservation Act) and resulting, ubiquitous dark, cold classrooms, people began to see conservation as synonymous with deprivation. The restrictions compounded management's energy problems. Not surprisingly, management was eager to distance itself from any association with such discomfort. The business office was being bombarded with strange jargon, such as kWh, Btu, and retrofit. The new lexicon drove financial people behind closed doors. Terms like non-recourse financing sent the techies to their boiler rooms—and even farther away from the upper echelons. Energy too often became someone else's problem. A subterranean group of us worried about it, but most folks just looked the other way.

Unfortunately, we let it happen back then. Even worse, it is still happening right now. In far too many cases, *the energy manager* is still buried in the boiler room and not at the management table where he or she could offer valuable guidance.

We are gradually letting an incredible opportunity slip through our fingers. Mentally, top management has relegated energy managers to the catacombs. And we have let them.

In our industry, the word "manager," as it pertains to energy, has become almost meaningless. How much **management** responsibility does the energy manager have in most organizations? How often does the energy manager sit at the management table?

I plead, "GUILTY!" I know I have not done my part. In fact, when allowed a little bragging rights, I am happy to tell people I am in the AEE Hall of Fame (not the Energy Managers' Hall of Fame). I quickly learned while working in DC to take clout when and where you can get it. And "AEE" has a lot more clout than "Energy Managers."

Can you visualize the energy manager climbing the corporate ladder to be the next CEO in your organization? Unfortunately, we can easily paraphrase that old song about not letting our kids grow up to be cowboys by cautioning them not to become energy managers. At present, there is not much future in it. It is a niche to nowhere.

The implications of having the energy manager in this dead-end job are huge. It has impacted our industry, our energy future, and even our political perceptions. Some corporations have circumvented this dilemma by giving a nod to the vice presidents for sustainability. This helps, but typically leaves the energy manager in a subordinate position. More tragically, those leading the sustainability effort are following the energy model revealing more "tech" than "manage." As a result, we are inclined to get more techies and too few *managers* addressing the critical aspects of sustainability. As noted in Chapter 6, sustainability management requires skilled leadership. The technology is important, but it is not enough. If we are to achieve our sustainability goals, we need to change behavior. That takes leadership. As we address the critical water component of sustainability, it will require truly challenging leadership, for some major behavior patterns will need to be changed.

With the limited recognition that energy managers have received over the years, it's a wonder we have the caliber of people we do occupying those key positions. We are incredibly fortunate that so many have served so well, but they clearly need a bridge out of the techie trap.

Before we worry about building a bridge for the energy manager, however, we need to get some idea of just how wide that chasm is.

Top management does not speak "boiler room." But then the "techies" seldom grasp concepts like discount rate, net present value and recourse financing. Energy is money. Burning energy we don't need is the same as burning money. But the *money* people in an organization and the *energy* people in an organization seldom talk. Put them together in a room and there is an awkward silence. The ills from the 1970s still prevail. Large segments of our energy community do not communicate with each other effectively.

Facility people, and energy people in particular, frequently express their frustration with trying to get management to understand *and appreciate* operation and maintenance problems. Over 30 years ago, I wrote an energy management manual for the Texas state energy office. One chapter in that manual, which spoke to business/boiler room communication barriers, was titled, "Is Anybody Listening?" It is a sad commentary on our industry (and maybe my writing) that the concerns in that chapter prevail today.

Energy is the lifeblood of our organizations, our economy, our national security, and our very way of life. It permeates every corner of our existence from the local elementary school to our most sophisticated industrial processing plant. It is an absolutely critical component of our life. But it typically does not get the attention it deserves. Unless prices go up suddenly or we have a 34-minute power outage delay in a Super Bowl game, few recognize the vital role energy plays in all we do.

The chasm is wide. The chasm is deep. All too many in our own industry don't really sense what a gigantic leap will be required to connect the energy managers of the world to the people in the business offices.

The bitter irony is that energy managers are frequently called upon to cut energy consumption, and that requires changes in occupant behavior. Unfortunately, too many energy managers lack the authority to foster behavior change. I beseech readers of this book to look where the energy manager is in your respective organization's chart. Contemplate for a moment how your organization works.

It's hard to effect change when a person is buried several layers below THE management. Occupants in a building, particularly operations and maintenance personnel, have a major impact on consumption. Clearly, they are critical to any attempt to reduce energy consumption. As we contemplate the organization chart, we might save ourselves a lot of grief if we try to envision how managing our absolutely essential water is going to work.

Instead of bridging the gap, we are too often prone to actions that make our problems worse. A lack of regard for energy within an organization takes on broader ramifications. There is a ripple effect. Many, many years ago, I was asked to serve as a guest professor at a university to discuss how energy policy is set in Washington, D.C. I was tempted to ask, "What policy?" Today, we have a policy that takes us away from energy efficiency and into "greener" pastures. Though our domestic production is much healthier today, we still merrily consume foreign oil while we are told to hate "big oil." In his book, *Why We Hate Big Oil,* John Hofmeister declares that oil and politics don't mix. In our current environment, he is right. Having rubbed elbows with all those folks in D.C. all those years, I would respond, however, that they *would* mix energy and politics if the public insisted on it. Politics and politicians do respond to the public. It's a little thing called being re-elected.

But looking around, we see energy neatly categorized down in the boiler room. It is someone else's problem. If it were **our** problem in the US, we would have a Keystone Pipeline underway.

In most cases, energy efficiency can more cost-effectively improve our environment than renewables. Yet, energy efficiency all too often takes a back seat to the exotic world of renewable and alternative energy solutions. If we only have limited amounts of money to invest in protecting our environment, we should be spending it where it will do the most good. Energy efficiency investments are more apt to leverage the money available. Energy efficiency can reduce pollution while it makes money. We can even use some of that money saved to buy down the cost of renewables. So why don't we? Why have the green proponents urged us to leave our energy efficiency needs wanting while we install more expensive solar panels? If we could get our act together, we could have both.

Much of the answer for this dilemma can be found in the level of regard that energy is given in an organization, and the focus of its *ignorance* can be found in the position the energy managers hold in our world.

Far too often, the energy manager's seat at the management table is empty. When strategic planning is done at that table, the voice of energy is too often silent. Everyone in our industry pays a price when energy concerns are held in such low regard. If we are to bridge the chasm, the first step is to get people to understand that it is a problem we all share.

The second step, is to recognize the role energy plays in our world and elevate the role of those who manage it.

While I was the Director of the Schools and Hospitals Conservation Division at the U.S. Department of Energy, we had a study conducted to assess the effectiveness of our Institutional Conservation Program (also referred to as the Schools and Hospitals Grants program). One important finding of that study was the vital role operations and maintenance (O&M) people play in our energy efforts. The study found that in an effective energy program nearly four out of every five dollars saved could be attributed to the energy efficient practices of the O&M personnel. The big dollar savings are not in the equipment, but in the people.

The evidence is before us. And has been for nearly 40 years! The same blind eye we turn on the value of the energy manager is the blind eye we turn on the O&M people. Lost in the upheaval are the financial savings that could be gained if only the energy efficiency practices of the energy manager were instituted and supported across the organization.

Energy service companies (ESCOs) often complain that they have to educate management to get the contract signed, and then they have to educate the O&M folks to make the project operate effectively. Why are these two totally different groups? Who at the management table can talk O&M? Who can convey energy needs to all the layers of an organization? And what are ESCOs doing from the outside to facilitate that communication and to build up the energy manager's role?

By now, we all know that performance contracts pivot on

the guarantees energy service companies make, and those guarantees rest on the risk management capabilities of the ESCO. One of the biggest prevailing ESCO risks is the customer. The greater the perceived customer risks, the more money the ESCO sets aside to hedge those risks. When that happens, both the customer and the ESCO have a smaller project relative to the size of the investment. Everybody loses. An attractive ESCO customer is one that is in tune with its energy needs, and that only happens when the energy manager is truly a *manager*. It is in the ESCOs' interests to use the project opportunity to leverage the position of the energy manager to a higher level in the organizational ranks.

If we are to elevate the concern for energy and the position that the energy manager holds, then that manager must play a key role. By his or her actions, the energy manager can command increasing respect and create reliance. This will only happen, however, if they do their homework and become the energy information resource the entire operation needs and can count on.

Until the day comes when energy managers take their rightful seats at the management tables, we all lose. Whether you sell energy efficiency equipment, energy services, or energy expertise, your job is made harder by the nearly total disregard top management has for the energy aspects of its operation. We all have a vested interest in enhancing the energy manager's position. But if we look around, we all have an even greater interest in the higher level of energy appreciation that goes along with a stronger regard assigned to the energy manager's job. This, in turn, will underscore the absolutely critical role energy efficiency should, and must, play in the health of our economy, our national security, and the planet.

Better yet, if we can finally smooth out the wrinkles in this energy model, it could pave the way to set up an effective model for managing water. There is much we still need to learn in valuing water and in deciding the quality of water used for specific purposes.

*We must not make the mistake of
putting the water manager on the sidelines.*

In the 40+ years since we had the great energy wake-up call, we have come a long way. Our knowledge of energy usage has evolved. But we still have far to go. There is a definite disconnect between the concerns we express for our environment and planet and the way we live. Those who espouse grave concerns do not seem to recognize the impact they themselves have as they jet to "eco" conferences. *We have a great propensity to indulge in circular finger pointing.* There seems to be a propensity to expect the other guy to be environmentally conscious while flying in private jets. Somehow we must shake the EROs (Environmentalist in Rhetoric Only) and focus on the job at hand.

There is much we could, and should, have learned about energy management by now. It would serve us well as we consider the pending, and in many locations immediate, water crisis. As we face our depleting resources, will our "lessons learned" be apparent? Do we have 40 years to get our water problems right? The answer is a resounding, "NO!" Our water crisis is at our doorstep; we just refuse to acknowledge it. And the longer we remain in denial, the more difficult and costly the remedies will be.

With minor modifications, much we learned from our energy experiences can be adapted to the absolutely critical water dilemma facing us. The parallels are striking. Our energy engineers have already walked that path. Our past "practice" on energy should be extremely helpful when we get down to the super-critical problem of managing water.

Chapter 3
It's Just Good Business

Theodore White once observed that the best way to influence someone is to sincerely ask for his opinion. The operative word in White's astute observation is "sincerely." When it comes to essential elements in our world, like energy and water, we can't afford to play games. To more effectively manage our essential elements, a critical component of the puzzle is human behavior change.

Getting people involved should be a fundamental part of any plan. The old adage about "psychological purchases" still holds. It's critical that people "buy in" when we are talking about something as basic as changing the way we use water. We may call it influence, persuasion, or just plain marketing.

The difficulties we had in selling energy efficiency offers fertile territory for ideas to sell the efficient use of water and the fundamentals of water management.

THE HARD SELL

Energy efficiency reduces operating costs, helps companies be more competitive, improves the economy, reduces pollutants, and *makes money while it's doing it*. It should be a no brainer. Yet in our work around the world, we found energy efficiency was a hard sell.

In many of the former communist countries, the answer was relatively simple. The economic system did not create conditions where the economic consequences of excessive consumption were apparent. There was a disconnect between the commodity and its cost. In fact, in countries like Czechoslovakia and Mongolia there was an incentive to burn fuel even when it wasn't needed. The system was based on how much one used this year. If you didn't use all your allotment this year, you'd get less next year. There was always the fear that next year might be colder.

Working with countries which were transitioning from communism to a market economy, I realized the commodity/cost considerations were crucial to their understanding how to select cost-effective efficiency measures. Finally, I hit on a way to get across the idea that wasting energy was the same as burning money. After spending two weeks in Hungary in the 1990s, I learned that the country imported three-fourths of its energy, but no one could tell me how much was being spent for that energy. At a point of great frustration, I took a match to a 10,000-forint note (about $37 today) and burned it, right in the middle of a presentation. They got the message. From Budapest to Beijing, the general response became a loud, "Oh, no!" And there was a growing level of surprise that energy and money were somehow connected.

In the industrialized nations, the reaction has been more subtle. Few seem to equate the relative cost of energy efficiency (EE) and renewable energy (RE) to determine the most cost-effective way to reduce pollution. In most instances, RE is the more expensive option, but the alternative energy people have done an exceptionally good public relations job of convincing corporations (and the public) that there is greater social responsibility in cutting fossil fuel consumption through solar, wind, etc. than with EE.

The beautiful thing about RE is that it can reduce our dependence on fossil fuels. It is increasingly obvious that we must do all we can to cut fossil fuel consumption. The question should be: "What is the most cost-effective way to do so?"

RE measures are typically more expensive per Btu saved than those saved through EE measures. But the options need not compete for budget allocations. Instead, the measures should complement each other to get the greatest return on the investment. It just makes great business sense to use EE when economically feasible and then use the EE savings to bring down the cost of the RE measures. A time will come, however, when the low hanging fruit of EE has been exhausted. At that point, the investment per Btu saved should be calculated for RE and EE and the most cost-effective measures pursued.

We certainly don't conduct our businesses in isolation. As with everything else, there are agendas within agendas. It's just good business to be aware of the factors at play. Budget items traditionally compete for funding. Outside forces, often out of our control, can determine the outcome. For example, the amount of support and encouragement you get from your utility will vary with its ownership. In the US, thanks to our federal government, the investor-owned utilities have less incentive to foster energy efficiency than the municipal-owned utilities do.

MAKING A DIFFERENCE

Conserving energy, be it through efficiency or the use of renewables, is a cornerstone of sustainability. It recognizes the value of using our resources as prudently as possible. Incentives to explore innovative ideas and increase the investment in alternative fuels can certainly be a plus. The end result is the ultimate in corporate social responsibility and just good business.

While the bottom line is not so obviously in play in the non-profit sector, the same thinking applies to public entities. In both cases, we all benefit from careful resource allocation, and so does the planet.

SELLING THE CONCEPT

Through the years, the drive for energy efficiency has grown, but as already noted, it has not been an easy sell. Despite its obvious benefits, energy efficiency has not been embraced by the business world as much as one would expect. Since this has been a universal dilemma, leaders in the industry around the world have puzzled over the problem. Aside from the money/commodity disconnect in the former communist economies already mentioned, companies in the industrial nations have also posed a problem.

Historically, in this relatively new industry most venders were focused on selling the equipment and the related energy savings. This posed a two-fold problem; first, came the awareness that *very few CEOs buy "energy"—they buy what it can do! Since the CEOs do not consciously buy energy, acquiring equipment and services that will cause them to consume less energy just does not seem to be on their radar.* At the bottom line, we weren't *selling* what people were interested in *buying.*

Secondly, the hardware is only part of the equation—and a relatively small part. As the industry moved from the traditional energy audit to the investment grade audit (IGA), it was found that the predicted results were more accurate if the IGA took into consideration the caliber of work performed by the operations and maintenance personnel. Since the IGA more accurately judged the effect of the people component as to how much energy was going to be saved, a closer examination of energy efficiency practices of the O&M staff seemed warranted. Once we became aware of the key role O&M played and the associated risks they posed, the predicted savings became more accurate.

O&M people might not sign the contracts, but they can exercise tremendous influence if they say "No!" Perhaps, more importantly, they are in a position to make sure the project succeeds—*or doesn't.* The O&M personnel are a vital component of a successful project. While facility people are getting attention from vendors, it is discouraging, even alarming, how often this hands-on market is still ignored.

Once the weakness in our marketing approach became apparent, I struggled to share this perspective. Then, a good friend told me that what we need to do is to learn to fish from the fishes' point of view. It works. Dangle the right bait, hook 'em and reel them in.

MANAGEMENT BAIT

More than one EE salesperson has bemoaned the difficulty of getting through to management about energy. Sales people will tell you that when the word "energy" is spoken, management's eyes glaze over and they rush to suggest who would be a good person in their organization to talk to. Chances are the sales folks are sent to the boiler room. For years, we deluded ourselves into thinking top management was uncomfortable talking about energy, but the bitter truth was, and is, *energy* is simply not on its radar.

So what interests them?

We have only to listen to management for a few minutes, and the most luscious bait we can put on the hook becomes obvious: money. Think for a minute of the last conversation you had with management. Do you remember any words about costs like: price, profit, bottom line, expenses? Or the more subtle, indirect reference to supply, availability, transportation, taxes, etc? It all translates to money

If you go into the head office and you talk about reducing operating costs without capex, their eyes will light up and you'll have their attention. If you talk about acquiring new equipment without increasing the budget, you'll get them leaning forward.

Then, you talk about finding money in their operation that is currently going to the utility for power they are wasting. Now that you have their attention, you can talk about ways to cut energy consumption (in general terms; save the tech stuff for operations folks) and use the savings to pay for the equipment.

A slightly smaller bait, but of the same persuasion, is to discuss competitive advantage. If an organization cuts operating costs using money already in the budget, it can cut product/ service costs and enable them to cut product costs to a figure below what the competition is charging. Now you are really speaking their language.

It pays to be aware when you wade into the budget arena, you could be going up against management's own pet project. Most likely, the person you are talking to got where he or she is by selling more product. Not surprisingly, there is a fundamental desire to do more of the same. They would much rather spend money on increased production, a new line of product, etc., than something as mundane as energy efficiency. Be prepared to walk them through the numbers. Table 3-1 offers an example of the comparative conditions surrounding energy efficiency and increasing production.

Table 3-1.
Making the Business Case for Energy Efficiency

	Energy Efficiency	*Increased Production*
Financing:	Avoided utility costs	New budget allocation
Effect:	Redirects existing funds	Add to budget
Impact:	Reduces operating costs	Increases operating costs
Product:	Priced more competitively	Apt to increase prices; less competitive
To realize:	Internal decision	Relies on outside events; Increased costs down line

In the end, energy efficiency can give them the resources to do more and sell more.

But making the business case in the front office is not enough; the facility people need to be sold, too. If the energy efficiency efforts are to work, the facility manager, director of maintenance, the supervisor of operations, etc. all have to be on board. For them, you need different bait. Fortunately the same bait that works for operations and maintenance (O&M) people will work up the facility line.

BAITING THE O&M HOOK

Getting to know the O&M people is the first step. Take time to see the problem from their point of view. Swim with the fishes.

Our painful history has shown that when utility costs go up (energy or water), the comptroller's office is most apt to cut the O&M budget.

At this point the facility manager, director of maintenance, etc. become their own worst enemy. They stretch the smaller budget to get the job done. At first the deferred maintenance doesn't show. The front office decides that the O&M budget was probably bigger than it needed to be and the cuts were okay, so when rates go up again, they slash again.

The facility people, who are feeling mighty underappreciated by now, are not going to see some "hotshot from outside" getting that coveted money for some gadget or service—especially when they were not consulted.

As with any lasting change in behavior, the trick is to become their partners—get on their side—speak their language. Explain how you will help get their message to the front office. Be sure they understand that the steps toward energy efficiency will help meet *their* needs.

If you are really good, you will carve out a small piece of the savings to meet some specific facility needs. Earmarking some savings to meet the needs O&M people have talked about is a sure way to have them become your partners. Instead of feeling threatened by someone telling the people upstairs what "should have been done," they will help make it happen.

Now, if you reread the paragraphs above and think *water* instead of *energy*, a parallel selling opportunity will emerge. As water prices go up, the same dilemma will play out, and very similar solutions will become obvious.

The one change I would urge is that we get organizations to think RESOURCE MANAGERS instead of energy managers. In most instances, we already have in place someone that with a little help can think water efficiency in the same way that they have addressed energy efficiency all these years. And with the change in title, it's a great time to give them a place at the management table.

MAKING PROGRESS

In the business world, we set goals. We list objectives. Then we check to see how we are doing. That's how we know when we are making progress.

At least that is the plan, but sometimes we promise the wrong thing. Sometimes we measure the wrong thing. And sometimes we don't measure at all.

In the world of energy performance contracting, the industry started out by not measuring at all. The contractors offered new equipment and a smaller utility bill. They delivered and everybody seemed happy—until the price of energy dropped and the energy service companies (ESCOs) could not meet their contract obligations. Then, ESCOs took another look and realized they promised to deliver on something they couldn't control. The contractors were promising lower utility bills, but somebody else was setting the price of energy. The ESCOs had very little control over what they had guaranteed.

Then, this neophyte industry shifted gears and realized the industry would be better off guaranteeing reduced *energy* consumption. The idea made good sense. Unfortunately, establishing documented results proved to be more complex than comparing utility bills. For example, if a facility was relamped and those old incandescent bulbs were removed, it changed the heating and cooling loads. Those incandescent bulbs were heat sources that just happened to give off light. Changing to fluorescent, or later T8s, increased both the heating and cooling bills. Calculating the energy savings was not quite as simple as we first thought.

The complexity was then compounded. The energy service providers, including ESCOs, told their customers, "This is how much we saved you." In one smooth move, they held out the other hand and said, "Now pay us." Gradually, an obvious conflict of interest began to shine through, and customers began questioning the process. The states of New Jersey and California sought to establish some kind of state-wide measurement standards.

But everyone seemed to have a different system. It's hard to compare results when the yardsticks don't match. Then, the feds did something right. US DOE stepped into the breach, gathered some experts and worked to come up with a yardstick that everyone could accept. The first attempt was a building energy management protocol, followed by a "North American" energy management protocol. In both cases, the measurement and verification (M&V) efforts were well received, but both lacked any kind of baseline guidance.

*It's hard to know how far you have progressed
if you don't know where you started.*

As mentioned in the previous chapter, the International Performance Measurement and Verification Protocol (IPMVP) eventually evolved. Ultimately, a baseline was incorporated in the guidance document. During this period, the IPMVP became a non-profit entity and joined with AEE to create an M&V certification program. With continued improvements to the IPMVP guidelines and the certification procedures in place, an internationally accepted standard was established. The IPMVP organization, doing business as the Energy Valuation Organization (EVO), continues to upgrade procedures and makes its latest version available through its website www.evo-world.org

The M&V guidelines have given backbone to the business case for energy efficiency. It is now possible to measure progress that meets both the owner's and contractor's needs. Developing a commonly accepted procedure to measure energy savings was a long, involved process. We are fortunate that work on water measurement has begun. The 2002 version of IPMVP, Volume 1 was even titled, "Concepts and Options for Determining Energy and Water Savings."

Now, we can all smile complacently because we already have water meters. Problem solved, right? No. When was the last time you had your water meters calibrated? Did you know that as the meters grow older, they lose accuracy and that the shift typically favors the supplier?

WHO'S REALLY RUNNING YOUR BUSINESS?

Many years ago, I had the good fortune to hear Mr. Vaclav Havel speak about the conditions in the Czech Republic. I found myself sitting there in Washington, D.C., envious of the freedom the Czech people (a formerly oppressed communist country) enjoyed. And that was before our more recent spat of regulations descended upon us.

When I was a child, doctors had private practices. Long before the Affordable Care Act, the amount of paperwork required of doctors was forcing them to pool their paper shuffling and combine their offices. It has only gotten worse.

You don't have to be a doctor to realize that regulations have changed the way we do business. Further, those regulations score a direct hit on the services and supplies we buy. When EPA regulations force the closing of a coal-fired power plant near us, the price of electricity goes up. It impacts others; so the cost of our supplies go up as well. Suddenly, the costs of doing business follow. Inevitably, the cost of our products and services must go up, too. About then, we worry if we'll take a hit in the market place.

For some of us who sell energy services, we may actually gain more customers as a consequence of EPA actions. We will suddenly find that many businesses in the coal-burning utility's service area are desperate to cut their reliance on electricity.

CHANGING THE WAY WE DO BUSINESS

They are rearranging the deck chairs on the *Titanic* again. For most of us, we'll get that sinking feeling when we look at our utility bills in the months ahead. If you are planning to expand or open a new plant, you need to assess the lay of the land. What is the local utility's primary fuel source? Is it about to change? Consulting engineers have more to worry about as their clients look around. The EPA is changing the landscape.

No question, the quality of air we breathe is a concern, but it has become almost impossible to predict the impact Washington, D.C., is going to have on your business in the days ahead.

The dynamics have changed. The good ol' "mom and pop" stores are increasingly a thing of the past. The challenge today is to do what we can to anticipate what the federal government is going to do and plan accordingly. *As resources become increasingly scarce, we can expect more government intervention in our business.* The prevailing sentiment in our nation's capital is that the folks inside the beltway know best, and they are there to protect you.

Despite burdensome regulations, coal is not going away anytime soon. But if the local utility has an old coal plant, your operation is in for some major changes. If it has a new coal plant, you are still in for changes. Under the EPA rulemaking, every state will be impacted differently. And in all its wisdom, the EPA forgot to really look at the regional implications. A thorough analysis of the new rules and *the state rules* that will be added requires important business precautions. From the business perspective, the net result of EPA's rulemaking over time in the US is apt to be the addition of many small nuclear plants. Not necessarily a bad thing, but another aspect of our power structure that will need more security attention.

The feds just can't resist "helping." As water becomes a scarce commodity in some localities, expect the feds to step in. The least of your concerns may be putting in wastewater storage and parallel plumbing for reuse water.

FEELING INSECURE?

If you are not feeling uneasy about your cyber situation, you are in denial. The need is paramount to secure our power supply. We are vulnerable. Those miles and miles of wire and pipes are out there lying naked, waiting for disaster to strike. The plants themselves are too often accessible. To the degree your business relies on power function, the more vulnerable you become. It should not take a tragedy like San Bernadino to make it clear that terrorism is very real and growing.

Decentralized power and micro-grids are becoming more and more attractive. At one time, having a big cogeneration plant on your campus was considered by many to be an eyesore. Today it is a thing of beauty.

Now, consider your water source. How well protected is it? How regularly is the quality of the water tested? What is the procedure for sharing the results of that testing? It only takes a spark from Flint (in this case Michigan) to have the feds decide the local people cannot effectively manage the water situation. The fact that the EPA knew since February 2015 that the Flint water was contaminated and told only officials (*not the public!*) will not change the "feds know best" posture. After all, EPA claimed it was working within the regulations. What do the regs say about standing idly by while our families, friends and neighbors drink contaminated water?

Too, the Flint situation was caused by decisions made locally. Consider if someone sets out to deliberately contaminate your water.

LOOKING AHEAD

Our businesses are increasingly embroiled in conditions and decisions over which we have very little control. To the extent that those conditions and decisions can remain local in nature, the more we will preserve some control of how we do business. Looking ahead, water is a *local* problem, not a national one. As water becomes a more obvious factor in how we plan our business operations, the more imperative it will be to preserve local control.

The longer we put off the planning for effective water management, the less control we are apt to have.

As we address water poverty, the business concerns related to energy that have been raised in this chapter will become more germane. Granted, water is relatively cheap now, and the connection still seems remote. Unfortunately, you can count on the cost of water going up. Emily Curley, sustainability manager at AtSite, offered an insightful glimpse into what is to come, "The End of an Era: Say Goodbye to Cheap Water." If you are following the indicators, you don't have to read her article to know she is right. The old supply-and-demand aspect will certainly be in play. But don't worry, you can count on the politicians, too. There is quite a group of them that see taxes as a way to control usage. Anyone hear the words "carbon tax" lately?

We are talking about major changes in behavior, which water poverty will demand. Getting the affected people to be engaged in the solution is critical.

Through the increased sensitivity to environmental concerns, the role of the resource manager should grow in importance. If management starts to see the energy manager in a broader context, folding in water management concerns will be a smoother fit. Those of us in the energy community need to start laying the groundwork now. The business case for effective water management needs to be made soon, as water poverty in one way or another threatens us all.

Chapter 4

Weighing Alternatives

As water supplies become more limited, what choices will we have? Will we begin to see the merits of reuse water, or desalinated water? Will we stop putting drinking water on the petunias? How will we react to higher prices of processed water when we all know we should be having only nice sparkling clear water available to us?

Now, *right now*, while we have the *domestic energy* at hand, is an excellent time to assess our options and to make realistic plans. Now, right now, while we still have *water* accessible to us, is an excellent time to assess our options and make realistic plans. Sounds like a stuck record, doesn't it? That's partly because they are both commodities—commodities that are critical to us.

Current conditions make it possible to examine our options without the fog of panic. When we buy energy, do we send any of those dollars to our enemies? Are the green lawns in Beverly Hills maintained at the expense of farmers in Northern California? Do we act like this situation can continue indefinitely, and we don't need to do anything about it?

Some of the energy source debates and the choices we have made through the years may answer some of those questions. So I offer a step back in time.

DREAMING WITH EYES WIDE OPEN

Strange dreams sometimes offer us food for thought. I recently dreamt I stepped into an elevator with rough hewn timber walls. It was natural and elegant at the same time. I had been told in my dream that this was the best way to get to the

5th floor.

The doors closed and the cage began to move so smoothly I could not tell if it was going up or down. I turned to push the 5th floor button, but there was no control panel, no emergency phone, nothing but rough hewn timbers all around me.

My "dream" elevator ride is frighteningly symbolic of our energy experiences. Based on faulty information, we are riding in a cage to parts unknown. We love the nature that surrounds us, but we don't know where we are, where we are going, and rather frighteningly we can't seem to control it. The greatest stumbling block we face is a level of obliviousness to the sources of the power we rely on. Even more frightening is our blindness to rising water needs and sources.

Aren't you even curious?
Don't you want to know if the next time you flip the switch
or twist the faucet, there will be power and water there?

Remember the oil filter commercial a few years ago that declared, "You can pay me now, or pay me later?" Remember that question was followed by an observation that *later* would cost more? We can start taking some of those critical measures now with a minimum of discomfort, or we can sail blissfully along and dump all this on our kids and grandkids. Let them pay the higher price tag.

Folks in the energy community are natural worriers. We are just built that way. But how well have we educated our colleagues and neighbors?

When we wake up from our dream, what floor will we be on? Will there still be time to do something about it?

Any attempt to get substantive information about essential commodities reminds me of another elevator situation. We were trapped in an elevator in a Cairo hotel. Seeking to report our problem, or get some information as to what was being done, we tried the emergency phone. It was dead. Once the situation was resolved, we told the hotel manager that the phone did not work. The next day the phone had been torn out of the elevator.

FILLING THE VACUUM

The absence of a coherent and comprehensive national energy policy, or even state policies, has been, and remains, strikingly obvious. The first step must be to take a serious assessment of what we have, followed by a bare bone cost/benefit analysis of our alternatives.

In retrospect, there were many times during the energy shortage when we desperately needed some hard-nosed number crunchers to tell us what it would really cost to transition to the next step. The sticker shock had us reeling from the price at the pump, and we too often failed to factor in the domino effect on all our other costs. If we had just ripped off the rose-colored glasses, we would have recognized that *every action pertaining to an essential commodity, like water, has a ripple effect* on every action connected to it. Every time we shift our gears, it has a sweeping effect on our economy.

With seemingly inexhaustible energy-consuming technological improvements and the richness of our ingenuity on one hand and borderline political malfeasance on the other, our energy horizon is foggy at best. We have been sitting on the edge of an energy precipice for decades, but we can't seem to gather our forces and treat our options with any inherent discipline. If we are to pull back from this precipice, we must realize that we critically need to be able to access *reliable* energy data. We need to break out of this pretend bubble that gives us false comfort. It is the kind of complacency that will lull us into a water catastrophe, and we'll be in over our heads before we realize what we've done.

Someone needs to burst our bubble—immediately. We cannot outguess the politicians. Even with good intentions, they too often act without giving sufficient attention to the unintended consequences. At the federal level, we have been constrained by the two-year window in the House of Representatives, a four-year window in the presidency, and a six-year window in the Senate. Statesmen worry about what's good for the country; politicians worry about getting re-elected. Given those conditions, most of our political leaders find it hard to do any long-term planning *especially if that planning reveals a short-term negative impact.*

Unfortunately, an overlay of uncertainties paralyzes our planning process. Out of this paralysis has come the hope of alternative energy. We launched a diligent search for fuel sources that were renewable and sustainable.

ALTERNATIVE ENERGY

A major source of our false comfort has been the belief that renewables will do the job. Alternative energy is important, it's attractive, and we should continue to pursue the power opportunities it offers. Several states have mandated an increase in alternative energy. Hawaii has legislated a goal of 100 percent renewables by 2045.

Incredible strides in solar, for example, have been made in India. An additional 2 megawatts are expected from solar just this year. The states of Tamil Nader and Telangana each contribute over 1100 megawatts now.

The growth of solar is depicted in the following illustration. The amount invested in solar has nearly tripled in five years. As it becomes more prevalent, prices will drop, and it will become a more attractive alternative.

Another promising source of alternative energy is wind, which is a green energy source and does not cause pollution. Some very sound reasons for using wind energy are:

- The potential of wind power is enormous—20 times more than what the entire human population needs.

U.S. Solar PV

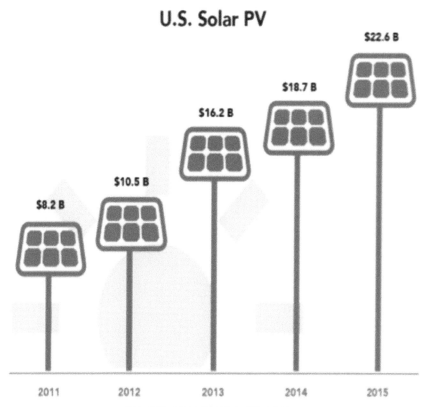

$22.6 B

$18.7 B

$16.2 B

$10.5 B

$8.2 B

| 2011 | 2012 | 2013 | 2014 | 2015 |

Source: Navigant Research for Advanced Energy Economy

- Wind power is renewable and there is no way we can run out of it (since wind energy originates from the sun).

- Wind power only accounts for about 2.5% of total world-wide electricity production, but is growing at a promising rate of 25% per year (2010).

- Prices have decreased over 80% since 1980 and are expected to keep decreasing.

- The operational costs associated with wind power are low.

There is, however, a negative side to considering reliance on wind. These concerns include:

- Upfront costs are high;

- It requires a lot of land to accommodate the turbines;

- Not all geographical and atmospheric locations are suitable for wind turbines;

- Winds are sporadic and back-up generation is needed;

- Getting the power from the wind field to the consumer is often difficult.

FUSION

A look at our energy resource options would not be complete without a nod to fusion. We've looked longingly at fusion for a long time. The carrot being held out to us has become all shriveled up.

The prospect is enticing, however, because it produces three to four times the power of nuclear fission. It produces no pollution. Any radio-active waste is minimal. There is no meltdown potential and its fuel is hydrogen. It's the answer to many of our energy problems, if only we can make it work.

Experts in the field tell us that the running joke about it being 30 years out is no longer true. In 2008, a ground-breaking development on a thermonuclear experimental reactor prompted large investments in further development. The plasma has been stabilized, but not yet at a high enough temperature. There is hope that a commercial level of production will be reached in about 10 years. There is even talk of its getting to the residential level in the 2040s.[4-1]

It's hard to sell a shriveled-up carrot. Fusion, however, could be the answer to our climate-change worries.

"DIRTY" RENEWABLES

Our love affair with renewables is compounded by an assumption that renewables are, by their very nature, low emitters of pollution, and non-renewables are high emitters of pollution. We tend to see "green" energy and "clean" energy as one and the same. This is not necessarily the case.

Fifty years ago, there was a tremendous push for wood-burning stoves for home heating. It was the *NATURAL* thing to do.

The trees would grow back, so wood burning was, and is, considered a renewable source. Ultimately, we began to realize that wood waste would not meet demand, replacement trees grow slowly, and the environmental benefit (sequestration) from trees could be lost. Much to the chagrin of some nature lovers, wood burning was finally recognized as a high pollution-emitting heat source. Today, for example, there are "burn bans" in the Pacific Northwest—a time, by law, when burning wood is not allowed because of the pollution potential.

THE CLEAN NON-RENEWABLE

Nuclear power is an excellent example of a non-renewable that is a low-emitting source. It is almost impossible to view the potential for nuclear energy objectively. Any consideration is clouded with concerns for safety and the association of this power source with nuclear armaments. This source of power, however, is relatively clean. The "smoke" coming out of the tower is all steam. Clean steam.

All of the nuclear power used in the U.S. goes to generate electricity. In fact, nuclear represents 19.4 percent of the electricity we generate annually.

Since new construction of nuclear plants has been relatively quiet for so long in the U.S., it often comes as a surprise to Americans that the U.S. has, at 30 percent, the greatest share of nuclear electric generation in the world. Other major shares of the world's nuclear power belong to France at 16 percent, Japan at 11 percent, Germany at 6 percent and Russia and South Korea at 5 percent each. Thirty-one states have commercial nuclear plants, the highest ranked being Illinois with 6 and Pennsylvania with 5. The best known is probably Three Mile Island in Pennsylvania because of an incident there many years ago. The story *not told* about Three Mile Island, however, is that when problems occurred, the back-up system worked.

If we were to ask around, many of our colleagues, who are not shy about expressing their preferences, will loudly proclaim the only solution is to rely on alternative fuels. A closer look will reveal a path liberally sprinkled with the joys of solar, wind, and bountiful non-polluting solutions. But in most cases that renewable road just won't get us all the way from here to there.

Discussions of pros and cons emerge for any source of energy, including fossil fuels and renewables. The bitter conclusion at this time is that neither solar, wind, nor any renewables as a whole are yet to be the panacea many would have us believe. It is immensely gratifying that renewables have been showing an annual increase around the world by 8.5 percent. Unfortunately, current figures indicate that renewables provide only 9.1 percent of our total energy supply—not nearly enough.

GETTING IT RIGHT

Energy credibility suffers from misleading statements and deliberate misdirection.

It is entirely too easy to point our finger at the other guy. Remember all the talk about the folks in East Anglia and "computer gate?" Or, the author of *An Inconvenient Truth*, who predicted rising waters and crumbling shores, buying a huge home on the California coastline? We shudder when we hear "plug in" cars don't use energy—and wonder where people think the juice comes from.

To meet the world's anticipated 2030 energy needs, current predictions say we will need 28 million megawatts—*28 trillion watts*. That's double our current usage. As we continue to assess these needs, we cannot help but worry about our carbon emissions. How much more can we absorb without doing irreparable harm to our planet?

In a very short 14 years we could have a demand gap of about 14 million megawatts—a gap equal to our current consumption. As staggering as this potential demand appears, even more daunting was the credibility gap that got us into this fix.

When we must use fossil fuels, they are there. Thanks to our infamous "Yankee ingenuity," we are enjoying an unexpected level of energy independence. We have gone from importing over 60 percent of our oil a few short years ago to now being able to flirt with exporting it. The benefit to our balance of trade is staggering. That balance and the jobs created in the energy industry and the general economy has created strong underpinnings to improve our economic outlook.

Our good fortune, however, is only a temporary hiatus.

We quickly forget how vulnerable we were just a few short years ago. We have also conveniently forgotten that our fossil fuels are finite. For all the problems associated with using that fossil fuel, we are not yet to the stage where we can do without it. But we could be using it more prudently.

Thanks to technology, we have an increasing range of alternate energy sources, and they are becoming increasingly cost-effective to use. The valuable strides in microgrid technology have made supply shopping and integration more accessible—and on a smaller scale. Essentially microgrids are Smart Grids in miniature, that can integrate distributed renewable generation, facilitate the use of electric vehicles, and take demand response to a new level of efficacy.

We are, however, in serious danger of being lulled into complacency. Our cyber world has put new demands on our energy supply. We are not only in peril from cyber warfare, we are also at risk of running short of the energy required to run all that technology. At the very least, all the new demands could raise the technology price tag to staggering levels.

The world is changing dramatically, and planning for tomorrow is fraught with unknowns. In a world increasingly relying on high tech, and the power to operate that technology, we are almost exponentially vulnerable to those unknowns. High on the list of uncertainties are how much energy will be needed, where we will need it, and in what form.

As beautiful as a green hillside can be from a distance, up close it can become a slippery slope. It's time we assess what we are doing and why. The global warming alarm hit hysterical proportions before we realized the scientific data were often shaky. Then, we moved to the term, "climate change," which is more defendable as climate has been changing for millions and millions of years. What we lack is sufficient data as to how much man's actions have to do with that change.

What we do know is that reducing carbon emissions can be a good thing. We know pollutants can have a negative effect on our health. Whether we are talking about the petrochemicals that go into making plastic, running around in our cars, or jetting to eco-conferences in our private jets, we must become more conscious of our actions and the impact those actions have on our environment, including our limited resources. Has anyone paused to consider all the bottled water we drink when we are at a conference lamenting our carbon footprint? We can't even argue that bottled water is better for us. Water out of the faucet is more fully regulated and monitored more closely than bottled water. Some figures on carbon related to the manufacture of those bottles would be interesting.

Recent findings have revealed that the folks in the "Black Triangle" in southern Poland have had a life expectancy that is roughly 8 years shorter than in other parts of Poland. Coal mining in the area is regularly named as the cause.

If we consider the multiple "Black Triangles" in one form or another in our world, we know we are doing some damage to our planet, which will cause long-term harm and may be irreversible. Action is needed, but we need to do a better job of weighing the unintended consequences of our actions.

The "clean energy" efforts, which regularly rotate through Congress represent good examples. Clean energy measures certainly have merit, but the potential economic ramifications of cap and trade, for example, are huge. Do we really want to find ourselves in a price tsunami that puts us at a competitive disadvantage in the world market? Competing countries are, of course, all in favor of our doing so. Expect loud applause from competing nations for any industrial cutbacks we make. In the long run, cap and trade could provide important environmental benefits, but what about the thousands more it will put out of

work? What about our settling for a lower standard of living?
How high a price are we willing to pay?

We have the time to get things in balance and plan for a
sustainable planet without extreme sacrifice. The question is: Is
a society which historically waits for a crisis before acting prepared to calmly weigh the alternatives? Can we take the time
to act responsibly as we seek to pass on a healthy planet to our
children, grandchildren, and future generations?

Can we rationally consider other measures we could take
that would yield roughly the same benefits without the economic upheaval? What about a much stronger focus on energy efficiency *plus* renewables? When will we become fully cognizant of
energy efficiency and demand response as sources of energy? It
would also help if we were to include relative cost-effectiveness
as a part of our planning.

If we take time to smell the roses, we also smell something
fishy in a proposed jump to 100 percent renewable energy. As
attractive as the picture may be painted and as impatient as

some may be, gigantic leaps do not appear to be part of our immediate future. We can applaud those corporations that declare they will be 100 percent renewable by a certain date, wish them well, and hope they don't price themselves out of business.

We must also examine our options along the way. It is critical that we assess our changing supply profile and our changing demands. It is imperative that we get from here to there in as cost-effective way as possible.

This all presupposes that we will have the freedom to call the shots. As we speculate about our energy future, *and our water future,* we will need to recognize that someone else may be holding the cards, the plans, and the final say.

Where is the renewable water? Are our hopes for renewable energy analogous to false hopes that we will find alternative water to meet our needs?

The elevator in my dream would not make it to the first floor... and somebody tore out the phone.

So where are we? And where do we go from here? After considerable research, the only conclusion one can reach is that we are one confused lot.

ALL IS NOT LOST

We are an innovative group. "Yankee ingenuity" is alive and well—even in countries that don't have "Yankees." After all, George Mitchell is credited with putting the pieces together to make fracking work in Fort Worth, Texas—not exactly Yankee territory.

We have only to look at the recent production from fracking to recognize promising technologies are still one of our most valuable resources. As an example, the potential offered by marine energy is particularly fascinating. With 71 percent of the earth's surface covered with water, that is a huge energy reservoir, especially when we consider that much of that water is constantly in motion—motion that might be milked as a power source.

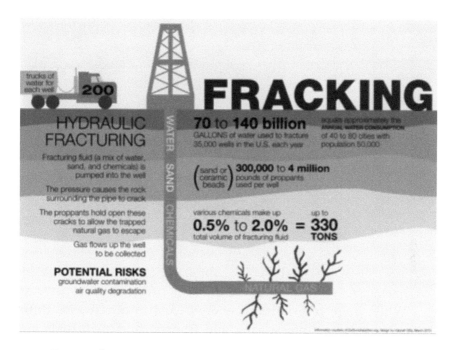

Research in refining old marine energy technologies and creating new ones is ongoing. These include wave power generation, tidal (current) stream technologies, salinity gradient power generation, and thermal gradient generation.

Each of these marine energy areas has a raft of energy capture ideas. For instance, in wave power generation, work goes forward in wave capture devices, shoreline devices, oscillating water columns, offshore wave energy converters, floats, wave pumps, etc. Exploring salinity gradient includes osmic power, hydrocratic generation, vapor compression and reverse electrodialysis.

As we pursue these opportunities, we must be mindful, however, that roughly 90 percent of life on earth is in the water. New species, especially those found at great depths, are being discovered every year. It is critical that the pursuit of marine energy be done in harmony with marine life. Small mistakes could cause serious damage as well as increase resistance to additional important research.

While huge reservoirs of energy exist in our world, diffi-culties with extracting, storing and transporting such fuels are often staggering. As an example, hydrogen, for all its abun-dance, poses major problems. Its advocates agree that resolv-ing hydrogen's extraction, storage and transport problems will require enormous research for decades to come.

Despite the opportunities marine energy, hydrogen and other alternative energy resources offer, we are not there yet. ALL alternative energy resources only contribute 9 percent to our energy supply. If we look at the most optimistic, but real prognostications, we find we do not have the resources or the time to meet short-term, or even long-term, energy needs with renewables. Maybe if we dusted off that shriveled carrot we'd find that the current surge in fusion development will pay off.

REALISM'S UGLY HEAD

Even with the best of hopes and lots of optimism, serious energy minds came to the ultimate conclusion with today's numbers that renewables are not going to fill the supply gap, and certainly not replace the fossil fuels currently being used. The disappointment this realization brings is compounded by the almost hysterical characterization of oil as being all bad. One of our problems is the role assigned to "big oil." We need to recognize that "big oil" in the U.S. is relatively small potatoes on the world stage. In proven reserves, our "giant," ExxonMobil, ranks 14th in the world. The ability of our "big oil" to set world energy prices has been vastly over-stated.

Another piece of our credibility gap is the "ANWR" problem. ANWR has become symbolic of our fear about drilling for oil—even on dry land. When pressed for details, most folks, who are dead set against drilling in ANWR, cannot tell you what the letters stand for, where ANWR is, how big it is, what percentage of the refuge would be involved in drilling, and what the "threatened" flora and fauna actually are. They seem unaware that we have been drilling for years right next door at Prudhoe Bay.

Ironically, those who seem most concerned about our planet, are far more willing to let other countries drill without the regulatory oversight we exercise domestically. They continue to drive their cars, air condition their homes, etc., without any real thought about where that fuel comes from, who is doing the drilling, or even what environmental damage is being done in far-off lands. We probably need to pause and remember that Earth is a relatively small planet, and the pollution created in those far-off lands could end up in our backyard.

THE GREATEST LESSON LEARNED

The greenest electron is the one not generated.
The answer is so simple. It is right there before us. It says it all.

Energy efficiency (EE) allows us to use less to achieve the same purpose. Energy efficiency means less energy needs to be generated to do specified work. When electrons are NOT generated, absolutely NO contaminates are emitted. Yet in the past decade, energy efficiency and conservation have taken a back seat to renewables.

As a supply option, it is well to remember that EE can produce revenue. While all other sources of energy will cost, EE can often pay for itself and even generate a positive cash flow. Even better, when paired with renewable energy, EE can help buy down the higher cost of renewables.

In our effort to reduce pollution, we sometimes overlook the fact that electrons not generated also represent an opportunity to conserve our finite resources. In our rush to find alternatives to oil, we tend to denigrate its value. To weigh all our supply options fairly, we need to recognize the unprecedented role oil has played in the evolution of the world we enjoy today. Certainly, the use of oil has presented us with environmental problems, but it has also served some valuable purposes. Between now and the time energy alternatives can meet all our energy needs, oil will continue to serve a critical, valued purpose. Energy efficiency can help preserve our resources to meet those needs.

For all its virtues, however, we must not lose sight of the fact that the best EE we can mount and the most cost-effective renewable we can field are still not expected to be enough to close the 2030 supply gap. Fracking has certainly helped close the gap, but we need to constantly evaluate our supply options. Sadly, all the strides we have made to use energy more efficiently in U.S. homes has been offset by the increasing technical demand for energy.

To make matters worse, the proportion of energy used as electricity in the US grew from 25 percent in 1978 to 42 percent in 2005. As we all know, our power plants typically require that we generate three times more than we actually deliver at the plug, so electricity consumes resources at a much greater rate. (U.S. DOE sites the Btu *source* value at 11,600 and the *end use* value at 3413.) This shift to a greater reliance on electricity gives us all the more reason to use energy more efficiently.

EE preserves our precious resources (a cornerstone of sustainability) and generates revenue while avoiding pollution. In summary, it is the best source of energy available. Wherever it is economically viable to use EE, it should be the preferred option in any energy management or sustainability program.

Renewable energy is an increasingly attractive source of power. Government incentives in many parts of the world are fostering more reliance on renewables, and by 2030 our reliance on alternative energy is expected to double. But even if it doubles, or triples, it will not be enough to close the supply gap, nor can we look to it to totally replace fossil fuels in our supply mix.

The data suggest that we will ultimately need new sources of supply. New supplies, however, require significant time and capital to design, build, install and distribute on any meaningful scale. It is inescapable that the search for more energy must begin with a search for more efficiency.

If we are to meet our future energy needs, we must boldly acknowledge that EE and RE will not be enough to meet our projected 2030 energy demands.

If we are faithful to our goals to preserve our planet *in its entirety*, the concern becomes one of who can explore, drill, transport, and use the oil with the least harm to our Earth.

In the last four decades we have been offered an incredible lesson in behavior modification. The advocates for green energy have been incredibly successful in selling the renewable concept.

We need to take a page out of the green biz world and a) further heighten the concerns about using fossil fuels, b) garner more support for alternative fuels, and c) *examine ways we can replicate this communications success when the water crisis swamps us.*

As you digest this chapter's brief glimpse at our energy situation and the choices we have, it becomes clear that we just may have been fortunate to have such a range of energy options. With water we have few. There is no alternative water. When we run short on potable water, we can turn to desalination, which will be relatively expensive. Or, we can reuse the wastewater, which may prove to be a hard sell at first. At present, no other options exist.

We have one more takeaway from our energy experience. It took us awhile, but we finally started weighing energy efficiency as a source option. We can do the same with water. We can use it more efficiently. That includes fixing the infrastructure; so we quit leaking one-sixth of our supply as we pipe it. Efficiency is a source we already have if we just learn to manage our supply better.

As we approach the world's water crisis, it is expected that more energy will be needed to run desalination plants. Fortunately, marine energy resources may be right at hand to meet the need. When we consider our water needs, associated energy demands, and a host of other claims on our resources, the growing need for sustainability management becomes even more apparent. The whole movement to use ALL our resources more judiciously and protect our planet has become a critical management issue. The *Sustainability Management Handbook*, from The Fairmont Press, offers a better sense of how these concerns can be managed effectively.

With over 40 years of weighing our energy options, we should be able to step back and see the big picture. We *should*, but can we? Do we know our end goal? Have we any idea of the most cost-effective way to get there?

As we continue to search for the best way to fuel our economy, there is much to be learned from the process itself. The implications, as we try to come to grips with the huge water crisis on the horizon, are significant. The benefits of fracking have some convinced our energy problems are over. A major rainfall in California will convince many that the water problems are over there as well.

These overnight solutions will not do the job. Looking back, we now know it would have helped to have had more information. We also learned a lot about resource transformation, and the critical need to use what we have as efficiently as possible. All of those lessons have real applicability as we face the coming water scarcity.

The 1930s Classroom

Amid the chalk dust and the lingering smell of wood oil cleaner of yesteryear, some valuable lessons were taught. Those 1930s classrooms helped lay the foundation for much we enjoy today. Along the way, however, mistakes were made. Since the 1970s, we have been on a steep learning curve.

The "Oil Embargo" of 1973 took us to school. It offered us a preview of things to come related to many of our precious commodities. The question remains: Did we learn our lesson?

Chapter 5

Troubling Tangents

A review of the indoor air quality hysteria associated with the energy efficient building offers us an excellent example of what could happen as we get farther into the difficulties associated with a water shortage. An outbreak of a communicable disease following a shift to reuse water in a community is apt to echo the indoor air quality frenzy.

As discussed in Chapter 10, Singapore, which imports all its water, reprocesses about 50 percent of it. As water becomes in short supply in certain localities, we are very apt to find parallel situations between the recirculated air and reprocessed water. Already we have headlines reading "How Much Poop Is in Your Drinking Water?" If any type of contagion is spread in an area where the water is reprocessed, the press will have a field day. The water version of "Sick Building Syndrome" will manifest itself overnight and grow exponentially.

Our problems with energy shortages seemed to send us reeling into an unknown land with problems at every turn. One that jumped out at us was the sudden belief that our indoor air quality (IAQ) problems had a huge cause-and-effect relationship to the energy efficient building. There is no question that the barricaded windows and the blocked air intakes were largely at fault in some facilities and responsible for part of the problem in others, but the relationship was frequently exaggerated in the press.

I think we can count on reuse water to give us another déjà vu all over again moment.

HOW TANGENTS CAN TAKE OVER

Much like dust devils swirl upward from a hot surface, major concerns seem to spawn associated problems. One of those dust devils evolving from our heated energy concerns in the late 1970s was IAQ.

Behind every dust devil, we seem to gather a juggernaut of jargon. "Tight Building Syndrome" became the phrase of the day. If we remember those boarded-up windows and air intakes, it is not surprising that nearly every IAQ article in the 1980s laid the blame on the energy efficient building. Not so surprising, in the rush to solutions the term "Tight Building Syndrome" was not only misleading but was inadequate in explaining the issue. The resulting focus was on ventilation and the related energy costs as the principle culprit.

With one eye on the utility bill, delayed start-ups and early shutdowns, which allowed the systems to "coast," became standard operating practice. Cracks were sealed. Variable Air Volume (VAV) systems became more common. The combined effect of these measures often caused the stagnation of air. They also put the systems out of balance and imposed part load operating conditions on the air handling and refrigeration systems. The ultimate, disastrous consequence from the IAQ perspective was the build-up of contaminants in the space. It took awhile before we began to appreciate the high price tag those "energy savings" brought about.

As we frantically sought solutions, the antithesis of the shutdown practices was ironically to increase ventilation. In this knee-jerk reaction era, air flow was not carefully studied. Too often, the ventilators were up near the ceiling. Increased air caused a nice breeze across the ceiling, but this short-circuiting did not necessarily improve the quality of air at the occupant level.

Many found reassurance in their rush to the ventilation solution when they looked to the air change standards. The American Society of Heating Refrigerating and Air-Conditioning Engineers (ASHRAE) had a standing committee (62) and its standard (62.1) is titled "Ventilation for Acceptable Indoor Air Quality." Obviously if one wants "acceptable indoor air quality," the answer was (and is) according to ASHRAE—ventilation.

Having served on the 62 committee, it can be firmly stated that any suggestion to change this misleading title met with strong resistance.

Into the IAQ vacuum rushed the media. Bad news travels fast and makes great headlines. This was certainly the case regarding our IAQ woes. Suddenly the pertinent trade publications, environmental magazines and TV network news were filled with information about radon, asbestos, PCBs, insecticides, lead, and mold. People became leery, apprehension increased, and "Sick Building Syndrome" was added to the lexicon. Stories were rampant. Teachers walked out of an urban school because of illness and odors. Public workers entered into a class action suit against the building owner attributing permanent disabilities to new carpeting. An entire public municipal building was abandoned because of humidity control and moisture problems. An entire school was closed down because it was discovered that there was some asbestos in a single floor tile, which had been drilled into. And the list went on and on.

Mr. Raum Emanuel, currently the mayor of Chicago, is famous for observing that politicians should never let a crisis go to waste. Many a politician rushed into the energy crisis and the IAQ crisis. Jurisdictional disputes arose among agencies as to who could claim IAQ. The Occupational Safety and Health Administration (OSHA) had claim to industrial IAQ and quickly wrapped its arms around other aspects of the problem. The Environmental Protection Agency (EPA) wanted a piece of the action and used the ASHRAE 62 committee to informally write some regulations.

Typically members of ASHRAE earn their stripes (and chairmanships) by serving diligently on various committees. EPA, however, found a shortcut by giving a tidy sum of money to ASHRAE, and suddenly an EPA government official was chairing the ASHRAE IAQ committee. Once the situation came to light, to ASHRAE's credit, the board voted that they wouldn't allow such practices and a new committee chair was found.

Our federal government seems to abhor a vacuum, too.

A glaringly obvious lesson from this experience is that when a critical resource becomes scarce and suddenly garners great public attention, one can expect the government agencies—and Congressional committees—to quickly jump into the fray.

As with the energy shortage, the IAQ situation gradually came under control. The single greatest step was the realization that an IAQ audit was possible and not necessarily expensive.

THE OTHER AUDIT

It quickly became apparent that in a walk-through (scoping) audit for energy purposes, an educated eye could identify up to 80 percent of the indoor air quality (IAQ) problems in a facility. With a little training, an energy auditor could do double duty. Simple remedies, such as drainage of condensate pans or changing inadequate air movement often cost little and could avoid major IAQ problems. (See *Managing Indoor Air Quality* by H.E. (Barney) Burroughs and Shirley J. Hansen, Fifth Edition.)

The trick is in recognizing the limitations of such an audit. To get beyond an 80 percent solution level, it typically takes extensive (and more expensive) testing. This additional work usually gets the problem identification to about the 90 percent level. Roughly 10 percent remains a very difficult challenge.

Further, the certainty in any of the findings is often nebulous. Companies that guarantee a predicted level of energy savings, such as an energy service company (ESCO), have been repeatedly cautioned to NEVER guarantee IAQ results. Inability to accurately measure contaminants (or the high cost of doing so) plus the number of variables that an outside company cannot control make guarantees impossible.

In our book, *Managing Indoor Air Quality,* Barney and I strongly recommend that if IAQ problems mushroom into full-scale concerns, owners and operators should not panic and rush to secure outside expertise. While it is important to respond promptly, a lot can be done through simple in-house procedures, and a few corrective actions may avoid calling for outside help entirely.

If it becomes necessary to retain the services of an IAQ diagnostic team, the preliminary work by the staff can greatly expedite the team's efforts. Cataloging occupant symptoms, their patterns by time and location, and recording the building conditions can prove helpful.

Health concerns should govern all actions. Should symptoms appear to be serious, life-threatening, or are likely to cause long-term health damage, a diagnostic team should be brought in immediately. The investigations should proceed as rapidly as possible. Care of the occupants should remain paramount, including the evacuation of the area or the building as appropriate.

Most facilities, especially those with potential health hazards, have a disaster plan in place. Evacuation procedures do not replace such plans.

Adding a water survey to the audit mix would not be very cumbersome and could provide the needed baseline data. Establishing the conditions that exist and the modifications that are warranted also provides a good defense when hysteria hits.

THE MANAGEMENT CHALLENGE

Isaac Asimov once observed that every industry creates verbal walls behind which fools cower. IAQ is no exception. Consultants are quick to embroider and embellish IAQ concerns and throw around a few Latin phrases to impress. Some just get caught up in their fields of expertise. Others are charlatans trying to cover their lack of expertise. Management should never hesitate to insist that the consultants talk in plain language. Scientists may take comfort in certain jargon, but managers need solutions.

After all these years, the best procedure is still the solutions-oriented approach. There is still much we don't know about detecting indoor air problems and there are still many difficulties associated with procedures to measure various contaminants. At first glance, it may not seem very satisfying, but the simplest process is to start with what we do know and chip away at it.

The most important aspects for management are to acknowledge the concern, respond appropriately, and communicate effectively.

There are two management concerns worthy of note here. The first is stressing that most IAQ problems can be traced back to inadequate operations and maintenance (O&M). In fact, this may be the greatest tie-in to our energy efficiency concerns. Put bluntly, too many managers pay the higher utility bills by cutting O&M budgets. That can come back to haunt the best intentioned folks. Second, because we pointed the finger at ventilation, it is only fair to identify the most common forms of inadequate ventilation, which may cause IAQ concerns. Many of the difficulties can be traced to facility management; so it is important to be aware of these factors. They are repeated below directly from the *Managing Indoor Air Quality* book. The reader who wants more detail is encouraged to refer to Chapter 4, "Investigating IAQ Problems," in that book.

- Not enough fresh outdoor air supplied to the office space;

- Poor air distribution and mixing which causes stratification, draftiness, and pressure differences between office spaces;

- Malfunction of system due to blocked or restricted make-up air louvers;

- Temperature and humidity extremes or fluctuations (sometimes caused by poor air distribution or faulty thermostats);

- Barriers to air flow and circulation from divider walls and working station partitions;

- Improper or inadequate maintenance to the building ventilation system.

- Inappropriate energy conservation measures: reducing infiltration and exfiltration; lowering thermostats or economizer cycles in winter, raising them in summer; eliminating humidification or dehumidification systems; and early afternoon shut-down and delayed morning start-up of the ventilation system; and

- Improper or inadequate maintenance to the building ventilation system.

THE TIES THAT BIND

If we do just a bit of digging, it is surprising the number of connections we can find among contributing factors. A bit of exploring related to the last bullet above will reveal a host of potential IAQ problems associated with improper or inadequate maintenance.

The bitter irony, and well worth repeating, is that many maintenance troubles are frequently the direct result of cutting the operations and maintenance (O&M) budget to pay the higher utility bill. It is not unusual to find utility line items as part of the O&M budget. Moving the dollars from one line to another in the same grouping is so easy.

In a survey done of the schools by the American Association of School Administrators in the late 1970s, administrators were asked where they found the money to pay the growing energy bills. Nearly 70 percent said the utility dollars came from the O&M budget. So when condensate pans are not drained and dirty filters are not replaced, it is not surprising that IAQ problems increase.

If we step back and look at some of these relationships, we find that many of our problems are traceable to management decisions. The problem is that we do not recognize the connection right away.

The good news is that an educated eye can spot a large majority of the IAQ remedies with just a casual walk-through of a facility. When it is paired with a walk-through energy audit, an auditor with some experience can often find the energy savings to pay for any IAQ work that needs doing.

In far too many instances the energy efficient building has been blamed as a cover-up for poor maintenance or to avoid the diligent search needed to identify the culprits. Improving IAQ while making provisions to use energy more efficiently just makes good sense. Those are integral parts of the building operations product that the facility manager is charged with delivering. Many factors which have a negative impact on energy efficiency also have an adverse effect on IAQ. Correcting such situations can improve energy usage and enhance the quality of the air. A few incidental measurements can tell quite a story. Consider:

- Poor maintenance of pulleys, belts, bearings, heating and cooling coils and other mechanical systems can increase resistance, causing a decrease in air supply. Good maintenance improves both energy efficiency and IAQ.

- Water-damaged insulation, ceiling tiles, rugs, and internal walls support biological growth. Wet materials nullify insulating properties. Correcting water source and replacement increases energy efficiency while removing sources of biological contaminants.

- Leaks in terminal boxes and valves reduce temperature control, cause occupant discomfort and waste energy.

- Inadequate or inappropriate service cycles of filtration can allow excessive pressure loss to build up and waste blower energy.

While energy and water needs cannot always be met with the same O&M actions, the few bullets above suggest some opportunities can do both, and they are not uncommon. There are many more that are worthy of a healthy inspection. O&M personnel should be trained to spot energy efficiency opportunities, IAQ needs, and water conservation opportunities as they walk through their buildings.

A review of the indoor air quality hysteria associated with the energy efficient building offers us an excellent example of what could happen as we get farther into the difficulties associated with water reuse. From the beginning in any water reuse, careful testing, hard data, and effective communications are essential.

Chapter 6
Sustaining Sustainability

Sustainability is a term rapidly swirling down the jargon drain. It is close behind *organic* as the overused and misused word of the decade. Remember when the world was neatly split between organic and inorganic and we really knew what organic chemistry was about? Now we have beautiful *organic* colors and the lady on the TV show, "House Hunters," exclaiming over an *organic* roof line. My favorite, however, was the kiosk in our local hardware store that bannered *"organic* seeds." Would someone please tell me what an inorganic seed looks like? Rocks?

While we have been intent on preserving our planet and its essential elements, some folks have come along and stolen the term "sustainability" right out from under us. The term is becoming a quagmire for those who want to sound noble and righteous. Unfortunately, we have a few corporations that have VPs for sustainability whose major responsibility is to make the company sound socially responsible. Too many have jacked up the term and slipped their favorite cause or agenda in under it.

To the extent that we let this happen, it undermines the original intent of the movement.

The United Nations 17 Sustainable Development Goals[6-1] shown on the following page, which undoubtedly had admirable intent at one time, seems to cover the waterfront. If we get the original idea of sustainability firmly in mind and then read the group's number-one goal—"End poverty in all its forms everywhere"—it is most confusing. There seems to be a classic case of misdirection. While lifting people out of poverty is a very laudable goal, it embraces a bitter irony. At best, one must wonder if the writers really thought this through. Consider: As people progress up the economic ladder, they typically consume more goods. Increased consumption tends to draw down resources and has a negative impact on the environment. In the history of man, the betterment of many has usually been done at the expense of the environment.

The long-term outcome of aiding people in poverty is not consistent with the effort to preserve the planet *unless* we change our perspective. First, we have to meld the betterment of mankind and the preservation of our planet into a common goal. To do this, we must first think of the human species as part of the environment. Then we must find a way to better mankind without doing it at the expense of the planet's essential resources. That might be the true challenge of sustainability.

1) End poverty in all its forms everywhere.

2) End hunger, achieve food security and improved nutrition, and promote sustainable agriculture.

3) Ensure healthy lives and promote wellbeing for all at all ages.

4) Ensure inclusive and equitable quality education and promote lifelong learning opportunities for all.

5) Achieve gender equality and empower all women and girls.

6) Ensure availability and sustainable management of water and sanitation for all.

7) Ensure access to affordable, reliable, sustainable and modern energy for all.

8) Promote sustained, inclusive and sustainable economic growth, full and productive employment, and decent work for all.

9) Build resilient infrastructure, promote inclusive and sustainable industrialization, and foster innovation.

10) Reduce inequality within and among countries.

11) Make cities and human settlements inclusive, safe, resilient and sustainable.

12) Ensure sustainable consumption and production patterns.

13) Take urgent action to combat climate change and its impacts (taking note of agreements made by the UNFCCC forum).

14) Conserve and sustainably use the oceans, seas and marine resources for sustainable development.

15) Protect, restore and promote sustainable use of terrestrial ecosystems, sustainably manage forests, combat desertification and halt and reverse land degradation, and halt biodiversity loss.

16) Promote peaceful and inclusive societies for sustainable development, provide access to justice for all and build effective, accountable and inclusive institutions at all levels.

17) Strengthen the means of implementation and revitalize the global partnership for sustainable development.

We also do some pretty weird things to preserve various species. Ever wonder what our life would be like today if we had protected the dinosaurs? Steven Spielberg has offered us a small glimpse. Whether one buys in on global warming, climate change or endangered species, it is time we quit grabbing the green paint brush and become responsible stewards of our planet. Instead of throwing thousands of loggers out of work and leaving their families destitute to protect the spotted owl, which by the way is doing much better today than the loggers, we need to really consider the consequences of our actions. Serious consideration needs to be given to the price we pay to "do good."

Speaking of species, there is another "sustainability" tangent that keeps getting in our way. We seem to have a penchant for confusing our concern for endangered species with sustainability. It is as if some of our colleagues see life on Earth in a freeze frame. Everything stops. In fact to hear some tell it, we should go back in time and preserve all that ever lived. We'd find ourselves right in the middle of "Jurassic World" with Rex as our pet dinosaur and pterodactyls circling overhead. Those flying lizards had wings up to 40 feet long. Wouldn't you love to have one of them perched in a tree in your backyard?

As we go forward, do we really want natural evolution to come to a screaming halt?

It's time we trim our sails and clearly state the true definition of sustainability. We desperately need to peel away some of the righteous palaver that is bogging down the original intent.

WILL THE REAL SUSTAINABILITY PLEASE STAND UP?

Back in the black-and-white era of early television, there was a show called "What's My Line?" After the imposters made several attempts to confuse the judges and the audience, the emcee would ask the "real" person to please stand up. If we are to sustain our sustainability efforts, it's about time we got the real sustainability to please stand up. We are about to lose our original intent in the mad shuffle that is going on.

Unfortunately, we have a multitude of self-ordained experts, who have declared what constitutes sustainability. At this point, I would plead with the masses to strip it down to the essential components. Maybe sustainability should be one step above survivability? Energy and water would definitely make the cut. What else would?

As the multitude of agendas crowd in under the sustainability umbrella, water concerns seem to lose out. Our water crisis must become a cornerstone of every sustainability plan.

As we contemplate sustainability, we begin to realize how many related issues and subcomponents there are. If we are to get our arms around it, we can add a few more ingredients that the "off the grid" proponents would want, such as waste management. The trick would be to agree on what constitutes the essential elements, then put a fence around it so that we have an integrated, unified approach. To give true substance and definition to sustainability, we need some consensus as to what a master plan should look like.

PICKING UP THE PIECES

In a recent paper, an author suggested that *maybe* we ought to include water as we address sustainability. Out of the goodness of my heart, I will not identify the source of this stupid comment. Suffice it to say that without water we will have NOTHING to sustain. I would direct this author's attention to the incredible amount of excitement out of NASA in 2015 when it was announced that evidence of water had been found on Mars. The excitement was founded in the simple fact that the existence of water indicated that there might be, or have been, some kind of life on Mars. If we turn this around, it means that if there were no water on Earth, there would be no life. Nada. Zilch.

The premise of this book is to look back over what we

learned during the energy shortage and see what might be applicable to an orderly process to address the coming water shortage. A quick look back reveals that after we stumbled around a bit, we learned in energy management that the first step is a quality audit. It is essential that we look around and decide what we have, what we truly need, and how well we are meeting those needs before we move forward. After we have built this framework, then we can indulge in the luxury of naming a few things we'd like to have.

An audit needs to be inextricably linked to a measurement and verification (M&V) protocol. The audit will establish an assessment of current conditions, ways to use what we must have more efficiently, and identify effective ways to sustain key elements of our world. It is only after a quality audit that we have a good read on existing conditions and sufficient information to most effectively state our sustainability development goals. The inclusion of M&V procedures will force us to list measureable goals and help us to formulate how we will gauge our progress as we work through the plan.

A sustainability audit is very similar to an investment grade energy audit (IGA). In fact, if you were to lay the two audits side-by-side, you would find striking similarities. This is deliberate. Our energy and water needs are crying for a more comprehensive sustainability audit. Transitioning to that should happen in a smooth, relatively effortless, and clearly logical way. Notice that the IGA is specified. After years of using the traditional audit, we became aware that more attention must be given to the people factor. Human behavior, attitudes and practices are critical components of an effective energy management plan. If anything, it is even more critical in a sustainability master plan.

The unifying theme here is the fact that in all cases we are assessing critical elements and exploring ways to use them more efficiently and effectively.

As the sustainability auditor becomes more sophisticated, the audit report will increasingly lay the groundwork for the master plan. The best audit in the world does not save resources or improve conditions unless it is put to use. At the risk of being redundant, I would like to take a moment here to underscore this point by repeating a quote from my *Manual for Intelligent Energy Services,*[6-2] which was also quoted in the *Sustainability Management Handbook.*[6-3]

> The audit is a valuable tool, but audits don't save energy, people do. The unattended audit report gathers dust. Only when it is read, discussed, and implemented can its energy/environmental/ dollar benefits be realized. The difference between dust and energy savings is people. It is the communication connection that makes it work.

The emphasis on people, their behavior, attitudes and practices reflected in this reference to energy audits should become the conceptual backbone of a sustainability audit, master plan and program. Technical components remain critical, but must be integrated with the responsibilities of the people who must implement the plan. The whole program should be held together with a very strong communications component. Behavior is more strongly affected if there is an immediacy to the supplied information; so, the effective use of social media should be woven into the effort.

The very essence of a master plan is the assumption that someone has "mastered" its contents and is running the show. As we move farther away from technical aspects and toward a focus on human behavior, leadership becomes increasingly important. It is impossible to overstate the need for leadership qualities in the person directing an organization's sustainability efforts. For the owner, it speaks to the need to maintain and replace, as warranted, resource-consuming equipment. As a sustaining critical part of the hardware assessments, however, there must be attention given to safety, health, codes/standards, compliance guidelines, operating expertise and efficiency. From the consultant's perspective, a master plan for the client is a value-added step beyond the audit. The planning component also establishes the framework for a long-term working relationship with the client.

If the audit/plan is going to serve its key function of modifying people's behavior, the people doing the auditing and planning must get into the field and listen to what the affected people have to say. In addition, the auditor must be constantly mindful of resource availability. As resources become more scarce, the cost of the material will typically go up; so costs as well as availability must be carefully considered.

Sometimes in our dedication to preserving a certain commodity, the economics are put to the side. When the cost/benefit component is ignored, there is a greater inclination for management to ignore the plan. Realism regarding costs is essential to getting top management buy-in. It is heartening, however, for owners and investors to know that if an investment is also designed to do good, it is more apt to be implemented.

For those wishing to conduct a sustainability audit and develop a master plan, you are strongly encouraged to read the *Sustainability Management Handbook,* and carefully examine the sustainability development plans and reports from a number of companies that have taken the lead in this critical area. Among those worthy of some contemplation are ConAgro, Steelcase, Walmart and Dow Chemical.

The audit and the plan must not only contemplate future endeavors, but also address what we do with the "old" parts as well. The demolition of Candlestick Park is a good case study and a reminder of the importance of managing waste.

CASE STUDY: CANDLESTICK PARK

For many San Francisco 49er fans, the demolition of Candlestick Park was like losing an old friend. As concrete crumbled, many fond memories of Joe Montana and other greats seemed to go with it.

But there is reason to take heart, for there has been a concerted effort to recycle much of the old stadium. The contractor, Lennar Urban, reports that it developed a "full demolition
recovery plan," which was approved by the San Francisco Department of Environment. The plan provided that 95-98 percent
of the materials will be recycled with approximately 92 percent
staying on-site to be used in the new structures.

Because of its prominence in the sports world, Candlestick
has become a high-profile demolition story. It offers an excellent
case study in the care of construction and demolition (C&D)
materials. C&D is no small problem. Nearly 500 million tons of
C&D material is generated each year. Despite the fragmented
and sometimes conflicting requirements in recycling, an average
of 70 percent is being reclaimed and C&D has grown into a $7.4
billion industry.

While the 49ers have had nothing to do with the actual
demolition, their new home, Levi Stadium, is LEED Gold certi-
fied and sets a standard for others to follow.

Candlestick Park

As we drive the proportion of recycled materials higher, there is a huge cut in waste and disposal concerns. At the same time, it unlocks much more economic value, which dovetails nicely with closed loop and circular economy models. It is the basis for a wonderful story about old materials and new revenues that can make the CEO's heart sing.

There is a sidebar story at Candlestick that must be told before we move on. The contractor, Lennar, came under attack after the Bay Area News Group published a story accusing the company of wasting thousands of gallons an hour of drinking water to hold down dust during the demolition. The story pointedly noted that recycled water was available less than 2 miles from the site (as part of a drought program). However, further investigation revealed that the contractor had sought to use the recycled water, but that state health codes prohibited the use of recycled water for dust mitigation. The restriction was designed to minimize the inhalation of potentially harmful substances when recycled water is aerosolized. This is a concern that must be addressed and resolved as the use of recycled water becomes more common.

A number of corporations now have strong sustainability programs. Others are headed in the right direction. If you have ever enjoyed Chef Boyardee's ravioli or Orville Redenbacher's popcorn, then you should rejoice in ConAgra's efforts to establish a four-pronged "climate resiliency approach." Their four "prongs" are:

1. Implementing energy efficiency strategies throughout its facilities to achieve a greenhouse-gas reduction goal of 20 percent per pound of product by 2020, drawing heavily on employee engagement and strategic capital investments in facility infrastructure, such as boiler control system upgrades, heat recovery projects and lighting retrofits.

2. Working with supply chain partners to assure sustainable, long-term sourcing of ingredients through sustainable agricultural practices and transport efficiency.

3.   Fine-tuning a corporate climate change policy, and

4.   Eliminating food waste in its facilities, thereby reducing the amount sent to landfills and curbing resulting GHG emissions.[6-4]

These are all admirable and desirable goals. Perhaps the greatest is "employee engagement" as that will be at the heart of changing the way we do business. If we are to sustain our efforts regarding sustainability, we must take a hard look at all aspects of our operations.

It is disappointing, however, that this statement from ConAgra does not explicitly speak to water management issues. The awareness in many corporations is simply not there yet.

As we search for guidance in using our resources more efficiently, new technologies can play an important part. One of the more exciting announcements in 2015 came from Dow Chemical's Water and Process Solutions group. Desalination and water reuse in industry has become one of the most important themes in the water sector in recent years. As part of Dow's sustainability goals, the company has developed a breakthrough technology in reverse osmosis. This product, DOW FILMTEC™ ECO, will be discussed more fully as we discuss remedies in the last chapter.

COP 21

The 2015 sustainability conference, COP 21, held in Paris did not fulfill the hopes and dreams of many in the world of sustainability. While attempting to put a good face on the results, the non-binding agreement coming out of the meeting fell short of the cited goals. On the positive side, the financial commitments and the shared city successes are expected to generate stronger support for sustainability in the months ahead. Materials being produced as a result of the conference should be rich in new ideas.

THE CORNERSTONE OF SUSTAINABILITY

Whether we are talking about lifting people out of poverty, conserving our energy resources, or preserving our environment, the primary goal of sustainability has got to be managing our water supply. Nothing else comes in even as a close second.

The prognosticators tell us we will run out of copper in 100 years. There have been rolling predictions on when we will run out of oil. It might be inconvenient, but we can live without either one.

We cannot live without water.

That is why this book devotes the last portion to building awareness of our dependency on water and strategies for more effectively managing what we have.

Chapter 7

It Only Takes Money

You want to reduce your energy costs; cut your energy use; minimize your greenhouse gas emissions; cut your maintenance costs; reduce some of the operating risks you face—AND AT THE SAME TIME improve your staff's comfort level; increase the value of your building and organization; upgrade your equipment; increase profitability and/or positive cash flow; demonstrate your commitment to your staff, your shareholders, your community, and the environment; engage your people productively in helping your organization grow (or perhaps a subset of those!). But of course, that costs money, and funds are hard to come by....[7-1]

Sandra McCardell

McCardell paints a great "want it all" energy picture. More importantly, she follows this quote with the question: Is there a solution? And then answers it with "No—there is not. There is not one solution … that fits all organizations, or all situations." But she does not leave us hanging, and adds that money can be found stating, "There are principles and strategies which are applicable to almost all organizations."

If we turn the McCardell quote into a check list, it rapidly becomes evident that we can't afford *not* to explore the opportunities. Chances are, as you consider your organization or a client, you will check most, if not all, of the following items:

- reduce your energy costs;
- cut your energy use;
- minimize your greenhouse gas emissions;
- cut your maintenance costs;
- reduce some of the operating risks you face;

- improve your staff's comfort level;
- increase the value of your building and business;
- upgrade your equipment;
- increase profitability and/or positive cash flow;
- demonstrate your commitment to your staff, your share-holders, your community, and the environment; and
- engage your people productively in helping your organi-zation.

It's a proven fact: energy efficiency (EE) makes money. Even better, energy efficiency is an investment, not an expense.

By now, it is probably self-evident that the concerns we face with energy are apt to be found in solving our water poverty as well. Financing water efficiency will be a challenge because it will be relatively new territory. The pathways to solutions, however, will be quite similar.

When we start considering energy efficiency as an opportunity to make money, the financing becomes more accessible. Under Technology in Chapter 12 are offered just two of many ways to make money in water efficiency. The opportunities are good, and in many localities, more urgent.

Traditional bank loans, GO bonds, utility programs, etc. are all available. There are many good sources of information devoted to the mechanics of financing energy efficiency. One excellent source is *How To Finance Energy Management Projects* by Eric Woodroof and Al Thumann.[7-2] Just think water and much of their advice will apply.

It should be noted that service companies need financing for the companies themselves; not just for the projects they undertake. For companies seeking venture capital, Red Mountain Insights periodically publishes a *Venture Capital for Energy Company Directory.* The ISBN number is 978-1-62484-038-8.[7-3] The folks listed in the directory are accustomed to gauging risks, and if you come to them with the right numbers and well-thought out plans, financing a company ready to do water projects just might be attractive to them.

The greatest source of funding energy projects has been the projects themselves. There is a whole industry out there that will guarantee that the savings can cover the cost of the project. For those ready to take the plunge, the companies that provide this service are called *energy service companies* (ESCOs). The industry is called Performance Contracting. There is a brief discussion of this financing approach later in the chapter. For more information, the first part of *World ESCO Outlook*[7-4] summarizes the current state of the industry and the second section offers brief separate summaries of the industry in 57 countries around the world. It is fertile territory for exploring overseas opportunities in the energy services and financing fields.

SIDE BENEFITS THAT MAKE CENTS

Before we discuss the advent of performance contracting, it makes sense to recognize that when big companies, such as Siemens and Johnson Controls, are ready to guarantee a given energy efficiency project can make enough money to pay for it-self, other funding avenues become cheaper and more accessible.

When we seek money, we are in essence renting it with the full understanding that we will pay it back. Banks are in the business of renting money. They need borrowers. But they want borrowers that will return the money with interest. In addition to the bank's obvious desire to earn a profit, the interest, which is the cost of renting money, is primarily determined by the inflation rate and the _perceived_ _level_ _of_ _risk._ Note the word "perceived." Bankers are in the business of weighing risks, and seeing the big boys put their money on the table significantly reduces the *perceived* risks. So whether an entity uses performance contracting or not, they can indirectly benefit from the people at Siemens who do energy efficiency financing. In fact, the whole energy industry benefits.

Improving the efficiency of energy use is recognized as the most cost-effective way, by far, to gain increased energy security, improve industrial profitability, assure greater competitiveness, and reduce the overall impact on climate change. We are all dependent on energy. It is the life blood of our economy, and recent advances in technology have increased that dependence. The more we are able to reduce our reliance on this outside resource, the better off we are.

Learning the most effective ways to reduce energy consumption has not come easily. Since the energy price shocks of the 1970s, we have grappled with ways to cut consumption. Through the years, we have gradually learned what measures most cost-effectively save energy. In the process, we have also looked for ways to get this information, and the financing it requires, to the end users.

The parallels to water efficiency are obvious and the business case for both of them is essentially the same. They are essential commodities and part of the cost of doing business. Cutting consumption of these commodities reduces operating costs.

SELLING THE BENEFITS

While the benefits of EE are clear, implementing them on a large scale has been difficult. Energy cost saving measures are technically and logistically diverse and often small in scope. They do not compete well for capital against capacity or market expansion.

From the financiers' perspective, high transaction costs and perceived risks make EE investments less attractive. In realizing these opportunities, it is very desirable to have an entity that can aggregate projects, demonstrate technical expertise, manage/ mitigate associated risks, and guarantee results. For those who understand performance contracting, that sounds very much like an energy service company, an ESCO. There are ESCOs that will do water projects. They are rather glibly referred to as WASHCOs.

The IEA's 2006 study, *Light's Labours Lost*, provides an excellent example of EE benefits. It found that should energy users install only efficient lamps, ballasts and controls, significant money could be saved over the life cycle of the lighting service. Such an investment would cause the global demand for electricity to drop substantially. In fact, the savings would give our increasing demand a free ride for quite a while. The lighting model would result in total demand remaining unchanged from 2005 to 2030. The bottom line: Avoided cost in total lighting expenditures would be USD 2.6 trillion and the avoided tons of emissions of CO_2 would be staggering.

The downside of this relamping surge, however, would be that it takes a huge source of quick paybacks off the table. It's harder to do the long paybacks if the quick paybacks have already been implemented and are not there to bring the aggregate costs down.

If relamping is done first, it may make it impossible for an entity to economically tackle bigger, less cost-effective measures, such as replacing a boiler or installing needed insulation. Saving the lighting measures gives the owner the opportunity to aggregate more measures, reduce the payback on the less cost-effective measures, and offer a bigger package to the investor.

Unfortunately, while putting off lighting work may make sense on one hand, we pay a price in delayed savings to do so. The cost of delay is a critical factor in determining what and when to engage in energy efficient improvements. We quickly get into a balancing act of weighing net present value, discount rates, and return-on-investment factors against each other.

Deciding what mechanical changes to make in the HVAC system is as much a financial decision as a technical one. That is one reason it is so tragic that many organizations have folks in the boiler room and in the business office who don't speak the same language or even talk to each other.

Finding an outside entity that can bridge this gap can offer multi-faceted benefits. The inherent risks in such endeavors for the outside entity, however, are significant. If we peal away the technology layers and some of the razzle-dazzle retrofit talk, we are back to energy service companies (ESCOs).

In energy performance contracting, specific consequences are guaranteed. The guarantee is based on the monetary value of the project savings, but it does so much more than that. Sounds like a wish list that would appeal to just about everybody, and it's totally doable for water as well.

The difficulties inherent to energy performance contracting (EPC) could have discouraged many embryonic efforts in this 1970s-80s fledgling industry. The absence of support, the absence of favorable legal frameworks, or limited financing constituted major barriers and could have defeated young and emerging ESCOs. Sometimes governments were there to help. Too often, they were major hindrances—either unwittingly or deliberately. When it comes to water efficiency, fortunately the battle has been fought, and the groundwork is in place. The Federal Energy Management Program office regularly refers to reducing water consumption.

It gradually became evident that ESCOs offered a good thing. The pathway to eliminating many of the traditional barriers for energy efficiency (EE) projects through reduced operating costs without capital expenditure gained traction. Add to that an industry that preserves precious human and natural resources and helps clean up our environment while making money. The story is a good one, and despite its problems, the ESCO industry has prevailed.

THE EMERGING ESCO INDUSTRY

Once it was demonstrated that EE makes money, ideas to capture this concept as a business proposition began to emerge. The first effort came from Scallop Thermal, a division of Royal Dutch Shell. Scallop took an idea, which had been used on the supply side of the meter for nearly 100 years, and gave it meaning on the demand side.

(CGC) had for years been guaranteeing savings from its work

in district heating. Now Scallop took this concept in the late 1970s to the UK and US offering to deliver conditioned space to its customers for 90 percent of their current utility bills. Scallop had determined a way to effectively manage facility conditions for less than the 90 percent baseline, and the concept of "shared savings" was born.

Ironically, both the UK and US governments initially opposed the performance contracting concept. In the UK, a little-known accounting officer named John Majors, later to serve as Prime Minister, created government language to permit such a program. In the US, an embryonic energy services industry got laws passed state-by-state to make it happen.

The struggles, however, to establish this young industry were far from over. The CGC concept, as developed by Scallop, provided that each party would share a predetermined percentage split of the energy financial savings. During the life of the contract, the ESCO expected its percentage of the cost savings to cover all the costs it had incurred, plus deliver a profit. This concept worked well, as long as energy prices stayed the same or escalated.

But in the mid 80s, oil prices dropped, and it took longer than expected for an ESCO to recover its costs. With markedly lower energy prices, paybacks became longer than some contracts. Firms could not meet their payments to suppliers or financial backers. ESCOs closed their doors, and in the process defaulted on their commitments to their shared savings partners. "Shared savings" was in trouble—and the process became tainted by lawsuits and suppliers' efforts to recoup some of their expenses. At the same time, facilities managers valiantly tried to explain losses previously guaranteed.

Fortunately, many ESCOs persisted in their efforts to make the new concept work. Some agreements continued to show savings benefits to both parties. Of even greater importance, several companies, which had guaranteed the savings, made good on those guarantees. Some even covered the guarantees of other faltering ESCOs.

FINANCIAL MODELS

Since it is expected that the ESCO model will embrace water efficiency more aggressively in the near future, it is important that our emerging water experts know the basics of the available options.

By the 1990s, the two dominant EPC models in the world were shared savings and guaranteed savings. With some minor modifications, the shared savings model is still common in Europe, and to a lesser extent in the North America. The economic viability of shared savings rests on the price of energy and the anticipated stability of that price.

After the oil prices dropped in the mid-1980s, ESCOs in North America shifted to guaranteeing the amount of *energy* that would be saved, and further guaranteed that the value of that energy would be sufficient to meet the customer's debt service obligations so long as the price of energy did not fall below a stipulated floor price. As energy prices dropped again in the second decade of the 21st century, the need for established floor prices was once more a critical contract component.

There is a second distinguishing characteristic in shared savings. In this case, as shown in Figure 7-1, the customer has no relationship with the financial institution and has little or no specific interest in seeing that the loan is repaid. Because all the savings must happen in the customer's facility and/or process, this factor further raises the risks to the ESCO and the financier.

Despite its price sensitivity, there are overriding reasons which encourage the shared savings model for the bankers, the customers and the ESCOs. Three major reasons are: the difficulty customers in transitional economies have in satisfying the bank's criteria for creditworthiness; a relatively new concept, such as EPC, is easier to establish in a country if the ESCO's customer does not have to incur debt; and the desire on the part of some energy end users to avoid incurring further debt, or to avoid going through the political/legal procedures to do so.

Shared savings, however, relies heavily on ESCO borrowing

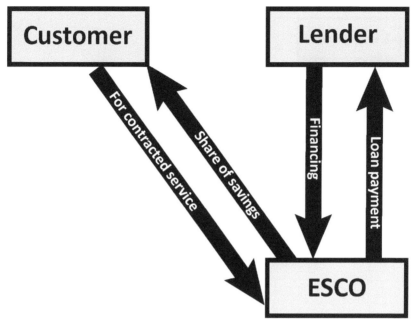

Figure 7-1.
Typical Shared Savings

capacities, and this presents a serious difficulty for small and even some big ESCOs to access financial resources rapidly or over time. After incurring debt on even a limited number of projects, an ESCO is apt to find that it is too highly leveraged to obtain financing for the implementation of more projects. This is a key factor in hampering industry growth.

To satisfy the ESCO's needs and to continue to avoid an untenable debt load, some ESCOs have turned to an emerging financial model which establishes Special Purpose Entities (SPEs), or in some countries they are called Special Purpose Vehicles (SPVs). The cash flow for this model is shown in Figure 7-2.

In this model, the SPV collects the revenues and pays the financier. Typically, the financial house and the ESCO are joint owners of the SPE.

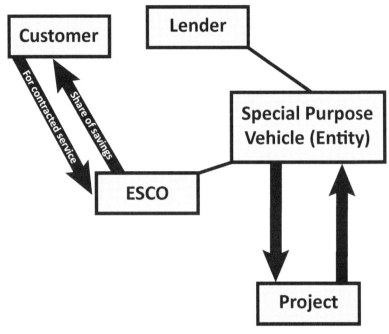

Figure 7-2. Special Purpose Entity

MOVING UP THE VALUE CHAIN

The dominant financial models are typically applied to comprehensive demand-side management (DSM) at the second level in the value chain as shown in Figure 7-3. Generally depicted at the next higher level in the value chain are efficiencies in supply, which might be district heating efficiencies, combined heat and power (cogeneration), or the implementation of stationary fuel cells. It is placed above comprehensive EE services only because the dollar amounts are typically greater for work on the supply side of the meter.

When, in addition to some demand-side services, the same ESCO provides a broader range of supply acquisition services, such as cogeneration or distributed generation, the package is referred to as an *integrated solution.*

The terms, integrated solutions and *Chauffage,* are sometimes used interchangeably, but Chauffage generally refers to a greater value-added approach. Integrated solutions may simply refer to a supply contract and a demand contract offered by the same ESCO, while Chauffage offers conditioned space at a specified price per square foot (or square meter). In such a case, the ESCO manages all supply and demand efficiencies. In practice, ESCOs sometimes focus only on supply efficiencies and refer to the contract as "Chauffage." It may include some type of ownership of a part, or the totality, of HVAC systems by the ESCO. The contract typically provides for some means of making adjustments for energy prices on an annual basis.

The ultimate value-added on the supply chain is the business solutions approach. Typically that approach allows an ESCO to propose solutions that make prudent business sense, which may go beyond reduced energy consumption. The ESCO may provide services beyond energy efficiencies, wherein the energy cost savings may help defray the costs of this additional work. In other instances, the work may actually increase energy costs, but lower the energy cost per unit of product through process efficiencies. The carpet factory case study by Duke Energy referred to in my earlier books on performance contracting

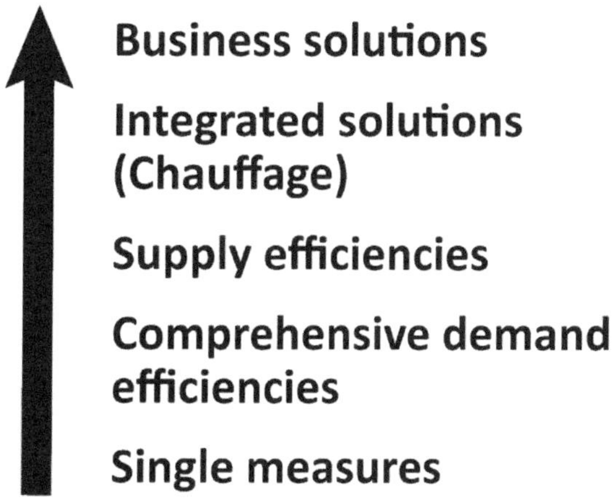

Business solutions

Integrated solutions (Chauffage)

Supply efficiencies

Comprehensive demand efficiencies

Single measures

Figure 7-3. The Value Chain

offers an excellent example of how this process might work. By changing the actual production process, the total energy costs went up, but production nearly doubled so the cost of energy per square yard went down.

THE ESCO APPEAL

Part of the attractiveness of EPC was, and is, in the models' ability to package services, equipment and related measures to create a project size more attractive to financiers. Another attractive aspect of EPC is the mechanism it provides for the ESCO to use *future* years of predicted energy cost savings *now*. For example, a project predicted to save $1 million per year on a 10-year contract can make $10 million immediately available for a project. Ten million dollars can buy a lot of services and equipment—and savings.

As EPC established itself as an accepted way to provide or enable financing for EE projects, it reinforced the efforts in Europe and inspired development in Asia. The ESCO industry began moving aggressively in all corners of the world. Soon, there were conferences designed to spread the word of effective best practices and to cement the existence of this new industry.

With the low energy prices in the second decade of this century, the guarantee floor price in contracts has once again emerged as a major concern. ESCOs, which did not insist on this protection are in trouble. The volatility of energy pricing remains a concern in predicting savings.

We are also feeling the pains of being hoist on our own petard. We did such a good job of selling energy efficiency that now owners have fallen into a pattern of assuming that the new energy-related equipment must pay for itself. Other items are replaced when they wear out. If an old boiler in a school quits operating in the middle of winter, the idea that the replacement should have paid for itself will not keep the children warm.

As water pricing becomes more of a factor, we can expect

volatility with that commodity, too. ESCOs and their customers will need to remember the shared savings lesson and guarantee the water saved—not the money.

Information gathering and comparisons of ESCO markets are limited by the fact that the notion of an *energy service company, or ESCO,* is understood differently from one country to another, and sometimes used differently by experts even in the same country. The problem with definitions has been highlighted in many forums and by numerous experts and business actors.

Put simply, an ESCO guarantees its energy savings performance. Parallel organizations, such as energy service providers (ESPs), may offer the same service, but do not offer the guaranteed results.

In any performance culture, actions are expected to have consequences. No matter what culture we are considering, both the prospective provider and the consumer need to carefully weigh the consequences. A cost/benefit analysis is key. Financing houses may provide this analysis for a client at no cost since the financiers need to study this for themselves anyway.

In the 21st century, the maturing of an industry is increasingly marked by the formation of industry associations. An affirmation of the growing ESCO industry came in 2007 when a networking meeting of the Asian country ESCO associations was held in Beijing just prior to the 2nd Asia ESCO Conference.

Once described as "alternative financing" by the US federal government, EPC is no longer an *alternative*; it is an accepted way of doing business. The ESCO industry delivers EE expertise, financing and ways to meet environmental mandates. The EPC concept is now pervasive and persistent around the world. But it has not happened overnight, and its development has been uneven. Many different barriers still limit the potential benefits that EPC could deliver.

The EPC model can be modified to do water projects. As water becomes more expensive and water efficiency becomes economically attractive, it is expected that water efficiency will become a routine part of the services offered by ESCOs.

THE WATER FIT

All of the ESCO models mentioned above, including the integrated solution, Chauffage, and business models, provide a business structure that will embrace water efficiencies. In fact, the water financial savings may exceed the energy savings and shorten the payback period. This was the case in a Boston University case study reported in an earlier book, *Performance Contracting: Expanding Horizons.*[7-6] At the university, a water resource management company, H$_2$O Matrix, joined with an ESCO to identify water saving opportunities. In this case:

*The inclusion of a water project actually reduced
the energy measures payback from 7.62 years to 5.6 years.*

As uneasiness grows amid a water shortage, it is well to remember that the mechanisms and expertise are in place to finance and service our growing water needs through the ESCO model.

Chapter 8

Energy and Politics

Energy and politics don't mix.
—John Hofmeister

We have been stirring the pot for decades. On occasion, we have added something like the Energy Emergency Conservation Act, which set school temperature restrictions—and stirred. Never content, we then threw in a little green energy—and stirred. Of course, the powers that be must get involved, so they decide to limit the use of coal—and stirred. In writing *Why We Hate the Oil Companies*,[8-1] Hofmeister was absolutely right, energy and politics don't mix. But that does not keep us from throwing energy and politics in the same pot over and over again.

The frightening thing is that after repeated attempts we have learned so little.

And the really scary thing is that our attempts to resolve energy concerns very likely foreshadow how we can expect politicians to resolve our water crisis. All your life you have heard that "oil and water don't mix." Try water and politics.

Hopefully, by the time you are done reading this book, you will be convinced that we have a huge water crisis on the horizon. I have asserted that our water crisis will make our energy problems seem like child's play. Let me see if I can make it stronger: Envision the "Clash of the Titans." Before us is a struggle that will make the Titans look like kindergartners at recess.

If we don't accept the inevitability of water poverty affecting us all directly or indirectly and, in the cool of the moment, develop some rational procedures for dealing with it, *it will very likely tear us apart.* In more than one discussion on this topic, leaders in the field are convinced that water resolutions will be the great leveler or wreak great havoc. Most bet on the latter.

We have had, and still have, some incredible opportunities to examine our energy problems and procedures and to find some helpful guidance in facing our water concerns. If the past is indeed prologue, we will need our politicians to act like statesmen and lead us to resolution.

So what have we learned in our energy struggles that will help us get ready? How do we get our politicians' attention and drive the point home?

If we are to make it work, we will need all the tools in the drawer. If we are going to avoid a really bloody mess and an incredible amount of pain, we should not try to drive a nail with our fist when we have a hammer at hand. We need to get out that hammer and figure the best way to use it ... NOW.

One of our best tools is knowledge. And right near the top of the list is knowledge about how our government works—or doesn't work.

Energy and politics may not mix, but that doesn't keep the politicians from trying.

WASHINGTON AT WORK

You don't have to be around Washington, D.C., long before you become aware of a basic philosophical struggle. It's the age old question as to the difference between a politician and a statesman. Cynically, but realistically, a politician has two goals: 1) get elected; and 2) get re-elected. In all fairness, if a candidate for office aspires to make a real difference, he/she knows they must first get elected.

In contrast, a statesman puts first what is good for the country.

In practical terms, most statesmen are apt to emerge when they no longer have to worry about getting re-elected.

If we contemplate the implications, the politician is less

likely to make tough decisions (defining "tough" as contro-versial, which translates into the fear of fewer votes.) So it is relatively difficult to find someone running for an elected office to declare, "If you elect me, I'll cut this program." Or, "I'll pass laws to cut back on these benefits you enjoy."

When you ask a politician to make a decision, there is a distinct possibility that he or she will be wrong—at least in some voters' eyes. So the temptation is great to put off the decision as long as possible. Then, to further obfuscate the situation, the legislative language is often deliberately made as vague as possible. In an effort to get a few of the wavering congressmen to come along, the language is further "softened." More than one law in Washington has been dubbed, "The Lawyers Full Employment Act" because the terminology was so vague that it was bound to be a litigious nightmare.

Those of you prone to reading legislative language may have shaken your heads and wondered why it wasn't more concise. You probably came away with a relatively low opinion of the legislative process. Be reassured. Those congressmen (or their staff) know what they are doing a surprising amount of the time. They are just crazy like a fox. Vague language gets more folks to sign on and to vote "yes."

When those decisions involve energy or water, we are ask-ing politicians to decide about something that is integral to the very way each and every one of their voters lives. The results will be invasive at best. A truly scary, job-threatening thing for an elected representative. Then, there is an inclination to make the language even more vague.

Unfortunately, for our legislators, energy and water do not lend themselves to vague language; so the inclination is to put off any action until we reach a crisis. The protracted actions on energy turned out to be uncomfortable but survivable. Wait-ing until things are desperate when it comes to water lays the groundwork for a huge catastrophe.

In the presence of such legislative hesitation, the celebrities or those seeking the limelight are prone to offer advice. While

Washington waffles, people will grab onto some poorly thought out ideas and run with them. We have already had William Shatner suggesting that the way to solve California's water problems is to run a pipeline from Seattle where "they have lots of water." However, we have been known to have droughts in the Pacific Northwest, too. The folks where I live are not going to be eager to ship water to Beverly Hills to they can water their lawns. Especially if it might mean the loss of jobs here.

When the crisis hits, DC seems to go into reverse. The hesitant can't wait to demonstrate that they are proactive on the "crucial issue." Politics suddenly rears its ugly head. The feeding frenzy is on and few worry about long-term consequences. If we consider having Seattle's water solve California's water problems, we need to start counting votes. How many more votes in Congress does California have than Washington state? Could the folks in D.C. order Seattle to ship water to California? Would the State of Washington comply? Could the situation escalate to the point that Washington might consider leaving the Union?

If the choice is between following DC mandates and local jobs, who knows what they will do in Olympia? Especially if one of those jobs happens to belong to a state senator.

Before it comes to that, some rational thought needs to be devoted to our water concerns. Again, the time to do it is before panic sets in. Our best leadership to guide us through this maze are those who brought us through the energy crisis. The obstacles are frighteningly similar and the inclination to act fast when we are in panic mode is always there.

HOW DO WE INFLUENCE THEM?

If the "them" are members of Congress, officials in the federal agencies, state officials and legislators, or local officials, I'd like to insert a short course, "Lobbying for Beginners," here.

The first step is to convince them that disaster is imminent, choices must be made, and then appeal to their statesman side. They all like to think of themselves as statesmen. Next, stress how critical the situation is and the value of making decisions *now*. And the argument should be loaded with poignant illustrations *from his or her state, district, or locality*. Remember when we were little and were told to clean our plate because children were starving in Europe? We need to think of children *right here in the US* who will be going without water and food.

One of my favorite quotes is from Theodore White when he wrote *In Search of History*, saying, that the best way to influence someone is to sincerely ask for their opinion. The underscore is mine. We can't just go through the motions. We need to care what they think, but to ask! They love it. It's really irresistible. Especially to folks in DC, who fancy themselves as anointed to take care of the "little people." (Yes, they really talk that way!)

This process is best conducted by the politicians' constituents; preferably someone who is regarded locally as a leader. Even better, see if you can find a local dignitary to give your politician a platform, such as city councilmen, the mayor, or the school superintendent, etc. Of course, it's imperative that they are fully and accurately informed first.

Fundamental to Lobbying for Beginners, it the basic understanding that we must give them the information they need to point with pride and/or view with alarm. They love pride and alarm because those make news, and visibility is crucial to a politician. Your job is to make it based on facts.

Ideally then, the discussion should include something that the politician can ultimately point to with pride themselves—the solution—or, a bad outcome that was avoided. Think picture on the front page or in a blog. Small kids are a plus. Bringing in a sympathetic press always helps.

Once a politician has bought into an idea and declares that a crisis is at hand, they will dig in their heels. If they shout it from the rooftops, the information must be RIGHT. If they are later proven wrong, there will be a mad scramble to save face. And the source of the bad information is not only the whipping boy, but is *persona non grata* from then on. So, while you are at it, it pays to give them a sense of what the opposition will be pushing. No political figure wants to be caught blind-sided. Of course, you round out the conversation by explaining why those other guys are wrong. If there is going to be a downside to the public, soften it and then stress how much worse the alternative will be.

There is one really big stickler when it comes to legislation designed to alleviate water stress: Members of the House of Representatives get elected every two years. They want to run on popular ideas. Water management language that forces voters to take short-term painful actions does not bode well for re-election. Most prefer to not vote at all rather than vote for a bill that will make voters unhappy. The *modus operendi* is to do nothing until the crisis forces their hand. Bills passed in a rush tend to cut with a wide swath, and individual or small group needs are often overlooked.

DOING YOUR HOMEWORK

Those who brought us through the energy crisis know how critical it is to get the right information, get it out there, and do it at the right time.

You can expect the opportunists and the misinformed to run rampant. It is critical to be better informed than they are and to be sensitive to the timing.

A few tricks of the trade may help. Get a committee in Congress to hold a hearing. Have one of those testifying ask their local press to cover the hearing. Nothing brings out the members of Congress like a bunch of cameras and lights in the hearing room. It's like watching moths circling a light bulb. There are junior grade staffers or interns whose sole responsibility is to sit in a hearing room and watch the preparation, so they can alert the members if something critical to our nation (like a bunch of TV cameras) is happening. The TV presence has the added benefit of underscoring the importance of what is being covered in the hearing.

Work with the committee staff. Get their help in having your people testify before 11:00 a.m., so the media will cover it that day. In fact, you can determine the chairman's position on an issue by checking to see who is scheduled before 11:00 a.m. The other side (if they make it at all) will be on the docket in the afternoon. The one exception is the spectacular opportunity to tear a witness to shreds. If a member can be assured that the witness will be decimated and the process will offer great camera fodder, then you will see them up there by 10:00 a.m.

If you can find data documenting problems that are about to emerge in the member's own district, that is even better.

Take full advantage of the social media to spread the word, but don't let it rule the day. Even in this day and age, the written word carries a lot of weight, especially if it is an original letter. A staffer once told me that his congressman had told him to count all the verbatim letters from one organization as one letter.

This all sounds rather Machiavellian, but we are talking about serious business with dire consequences. And what it takes to get Congress and political leaders to act is apt to be of extreme importance.

Photos have impact. If you have something that looks like "front page stuff" do not hesitate to make it available to the Congressional staffers. Consider a small child scooping up some murky water out of a stream with a heart-breaking woebegone expression on her face. Do you know how many children die each year from drinking contaminated water? Get the current facts and pass them along. Then the message is: Doing nothing is tantamount to letting the poor children die.

JUST GETTING STARTED

Even if we get a law passed authorizing some much need-ed action, our work has just begun. Implementing a law takes people and money. This means that the responsible agency must have money in its coffers to make it happen. We still need the appropriations folks to allocate some money and the budget folks to include it in the final budget. This is why the Committee for Full Funding in the education community used to work so diligently to get authorized programs funded.

But even that is not enough. Once the needed legislation is enacted, the responsible agency needs to write the regulations. Because the laws are deliberately written vague, the regulations become absolutely critical. For decades the guidance to lobbyists has been: "If you have a choice between the law and the regs, take the regs."

It is imperative that advocates find out who is responsible for writing the regs and then *gently* educate them. It is well to keep in mind that people writing regulations have a lot of power—and typically the ego to go with it—so respect them. Be helpful. Offer quality information. But let them do their job.

After the team has written the draft regs, folks up the chain of command have to bless them. But we are not done yet. The Office of Management and Budget has the final say before they are published in the Federal Register.

This gives you a rough idea of the hoops that need to be jumped through. Some procedures change with time, administrations, and specific agencies. But one thing is constant, a lot of people need to have a hand in every effort, and chances are they will not fully agree on what needs to be done.

Those in the energy industry can attest to the complexity and confusion that has surrounded energy legislation. Unfortunately, the actions needed to address our water crisis are apt to be far more complicated.

One other caveat, water poverty is most likely to be a local problem. Typically, it will be best served with local solutions. But once the feds get their hands on a problem, it changes the scope and often the focus. What is decided in DC may be totally inappropriate to meet local needs. The Message #1 is: Solve what you can locally, and do it as soon as possible. Message

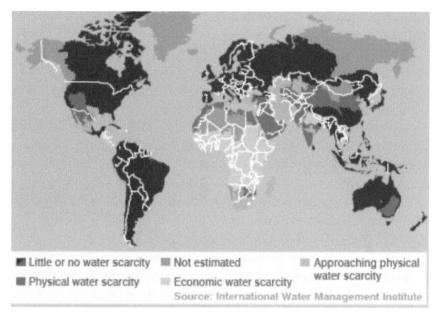

Figure 8-1.
Physical water scarcity and economic water scarcity by country. 2006

#2 is: If it gets to the DC level, go back and read this chapter again.

Water scarcity is the lack of sufficient available water resources to meet *water needs* within a region. It affects every continent and around 2.8 billion people around the world at least one month out of every year. More than 1.2 billion people lack access to clean drinking water.[1]

Water scarcity involves **water stress, water shortage** or deficits, and **water crisis**. The relatively new concept of *water stress* is difficulty in obtaining sources of fresh water for use during a period of time; it may result in further depletion and deterioration of available water resources. Water scarcity can result from two mechanisms:

- physical (absolute) water scarcity
- economic water scarcity

Water scarcity is expected to be the source of much discord in the future.

Ismail Serageldin, former vice president of the World Bank, warned over 20 years ago, "If the wars of this century were fought over oil, the wars of the next century will be fought over water."[8-2]

Water is also being referred to as "blue gold." The documentary, "Blue Gold: World Water Wars," gives an opening glimpse of the kinds of international conflict that could be on the horizon.

Water is a national strategic resource. Fights over riparian water rights are apt to increase. The term for water politics has already been coined—hydropolitics.

Others have argued that water interests transcend national boundaries and nations actually stand to gain from cooperative efforts regarding water supply issues.

INTERNATIONAL RELATIONS

So far we have addressed domestic problems and related political implications. Water scarcity is destined to become a huge factor in the relations between countries in the 21st century.

On its website addressing the global water outlook, the UN states: "Water scarcity is both a natural and a human-made phenomenon. There is enough freshwater on the planet for six billion people but it is distributed unevenly and too much of it is wasted, polluted and unsustainably managed."

The UN goes on to conclude: "Water scarcity is among the main problems to be faced by many societies and the World in the XXIst century. Water use has been growing at more than twice the rate of population increase in the last century, and, although there is no global water scarcity as such, an increasing number of regions are chronically short of water." [underscore supplied]

It is important to recognize that water scarcity is a wandering specter. It is not likely to become a "global" problem, but there is always drought somewhere. It seems to fall on the unsuspecting and create personal and economic havoc as it goes.

Finding a reliable source of safe water is often time consuming and expensive in the developing world. This is known as economic scarcity. Often the water can be found, but it takes the economic sources and the political will to do it. Economic water scarcity is a very disturbing form of water scarcity because it is so often rooted in a lack of compassion and usually reveals a lack of good governance, which allows the condition to persist.

More than one in every six people in the world do not have access to potable water. These regions are described as water-stressed. One measure of water stress, the Falkenmark Water Stress Indicator, defines a country or region as experiencing "water stress" when annual water supplies drop below 1,700 cubic metres per person per year. Limited water shortages are expected in regions with 1,700 and 1,000 cubic meters per person per year. At below 1,000 cubic meters per person per year, the country is described as facing water scarcity.

Figures from 2006 estimate that 700 million people in 43 countries were living below the 1,700 cubic metres per person threshold. Water stress is ever intensifying in regions such as China, India, and Sub-Saharan Africa, which contains the largest number of water-stressed countries of any region with almost one fourth of the population living in a water-stressed country. Ironically, the region of the world so rich in oil, the Middle East, is one of the world's most water-stressed regions, with averages of 1,200 cubic meters of water per person. In China, more than 538 million people are living in water-stressed regions.

All of these statistics and the findings cited in Chapter 11 do not bode well for world peace. The political ramifications are huge. The challenge to world leaders will be immense.

When a commodity, like water, is vital to human life, critical to economic welfare, and a strategic resource, political involvement is inevitable. Our water experts will be equally challenged to make sure those leaders act in an informed manner.

Chapter 9

FED UP: Life Inside DOE

Nearly every newspaper or blog story related to some federal government action laments how "Fed Up" we are with our federal government. It's almost a parlor game in Washington, DC, to bemoan ways that we could do it better. When I was the head lobbyist for our nation's school superintendents, there were many days that I was "fed up."

But our critiques from outside the federal fortress, frequently do not fit the inner sanctum. It is a real eye-opener to get inside the government and see how it actually works. There is an old line about the ugliness with which sausage and laws are made. It was undoubtedly first professed by someone who had made laws.

First of all, there were, and are, some great people working for our federal government. The devotion to our nation and the hours that some work are incredible. Unfortunately, there are, some who are just putting in their time. Even worse, there are some who feel THEY have all the answers and have a great obligation to take care of the "little people."

It's a fascinating place. I'm glad I worked there, and doubly glad I didn't do it for long.

It didn't take long to realize that what our government does is pervasive and effects our day-to-day actions. It became equally key to realize that affecting changes, or managing a crisis, must involve the government. The art of persuasion in DC has some special twists. To be persuasive requires that we get a basic grounding in how our government works and how it "thinks." It looks much different from the inside.

THE OFFER

I was at Disney Land (rather fitting in retrospect) when I got a call from the United States Department of Energy (DOE) inquiring if I would like to serve as the head of the Office for Schools and Hospitals Conservation. After some cogitation and negotiations, I agreed to take the position with the understanding that I could head up the task force to revise the regulations governing the Schools and Hospitals Energy Grants Program, more popularly known later as the Institutional Conservation Program.

I had complained rather loudly and frequently about the initial operation of the program. My husband always proclaimed that the folks at DOE offered/gave me the job to shut me up. He may have been right.

The first signal I got that life "inside" might be different than expected came when I reported for work at the U.S. Department of Energy. With a great sense of purpose. I arrived to "save the day." It took about 30 seconds to realize that I was not about to save anything.

There was no welcome. No introduction to how things were to be done. No office. It quickly became apparent that the job I had been offered was no longer available. It had been taken by the man who had originally offered me the position. It seemed while I was vacationing in Europe between jobs, there was a political "realignment" at DOE and my position became an attractive "bolt hole" for the boss.

The US Department of Energy is located in the Forrestal Building, pictured in Figure 9-1. It is impossible to go from here to there. If you are on the fifth floor in one part and want to go to the sixth floor, you must first go down to the lobby. It is incredibly easy to hide out in the place. My new boss operated in an avoidance pattern for two days. He really did not want to face me. Ultimately, the word came down. I got a *desk*, but would share an office and a secretary. My title would be Director of the Schools and Hospitals Energy Conservation Division.

Figure 9-1. The Forrestal Building, Washington, D.C.

I was left wondering what my responsibilities would be. Such are the ways of the bureaucracy.

Speaking of bureaucracies, one aspect which typically goes unreported is the incentives built into the system. A good example is the way the government is organized. If workers are not "Schedule C" (political appointees), they are paid according to their government service (GS) level. A major component of the GS is the number of people you have working for you. Seems reasonable until you start questioning how one goes about getting a raise. Want higher pay? Then, get a higher grade. Want a higher grade? Then, get more people to work for you. In large measure, it is a numbers game.

In management jargon, there is also a thing called "span of control." The broader the reach, the more people, the higher pay grade. Whether it is relevant or compatible with current responsibilities is not necessarily a factor. The major criteria is to get your arms around it. Not surprisingly, this fosters on-going turf warfare.

A good example is the waves caused by the Environmental Protection Agency (EPA) regarding asbestos in the schools. When the button was first pushed, the issue belonged to the Occupational Safety and Health Administration (OSHA) in the Department of Labor. I was told by a relatively well placed EPA employee that the higher-ups wanted to get their arms around asbestos in the schools issue and he wanted a higher pay grade; i.e., bigger department. They saw this as a real bonanza. Through some clever maneuvering, the agency got itself sued and before we knew it, asbestos in the schools became a big issue.

Citing a toxic substances control act (TOSCA), section 6, the agency started its rule making procedures. Unfortunately for EPA, the law required the agency to produce data it did not have. Ultimately, the agency got another law (AHERA) passed. Despite the gazillion ads for mesothelioma on TV, the evidence, according to The Health Effects Institute and other reliable studies, is still questionable. The bad news is that the whole thing cost the public schools (and the taxpayers) billions of dollars. Further bad news is that all that money could have been used to help educate kids. The good news (I guess) was that the EPA guy got his pay raise.

GAINING A WORKING KNOWLEDGE

Some of the best going-to-DC advice I got was from a professor when I was working on my doctorate. When he heard I was headed for Washington, he told me that at anytime I was in the hallways, even when going to the restroom, I should carry a manila folder (this was before iPads) and walk purposefully. The underlying guidance was to act like you know what you are doing. It works! I would add to that: Be friendly and listen a lot.

If you want to take the passive route and not make waves, you can last for years. If you are dedicated to making a difference, you soon learn how the square peg feels when shoved into a round hole.

But active listening helps. Through the grapevine, I soon learned that a task force to revise the grants program regulations was being formed. The friendly word was that a colleague, Harry, assumed he was going to head up this task force. Without getting into the sausage thing again, I am happy to report that my boss honored his promise and I did head up the task force to promulgate the revised regulations.

What a fascinating process! It became readily apparent that many decisions with far reaching consequences are made on the back of the proverbial envelope. There is a wonderful mix between "We, of course, know better," and "Ignorance is bliss."

In all fairness, however, it should be noted that this was 1980, and there was an incredible amount about energy conservation and efficiency that we did not know. For example, one of our challenges was to assign a Btu value to the end use kilowatt hour. After a number of inquiries, I called Don Carter, who at the time chaired the energy committee for the consulting engineers group, and he suggested it was probably 3412 or 3414; so I said, "Let's make it 3413." So we put it in the regs, and 3413 was cast in bronze.

In this world, many decisions are made using carefully developed scientific research. Many are not. For example, ever wonder how the decision was made that 15 cfm was the right amount of air exchange in a hospital? One day I was visiting with a couple of engineers at HHS (Department of Health and Human Services) and they explained how that magical number was derived. At an earlier time, they apparently had sat in a room observing some condensate on the ceiling in a corner of the room and decided to calculate what it would take to remove it. Yep! Their answer was 15 cubic feet per minute. Once the good engineers realized they had let the cat out of the bag, they noted that as a member of a "sister agency" what they had shared was, of course, in confidence and I should never share it. Right!

Naturally, nothing goes as planned outside—or inside—the government. More steps and more people increase the likelihood that things will go awry. The feds are undoubtedly the epitome of that. When the task force completed the draft regulatory revisions, the draft regs had to go through mandatory concurrence—a bureaucratic procedure for everyone signing off up the chain. Then, the draft had to be reviewed by OMB (the Office of Management and Budget). Finally, they were ready for publication in the Federal Register. And it was at that point the new president of the United States (POTUS) froze all rule-making procedures.

When they were "unfroze," those lovingly developed regs had to again go through mandatory concurrence under the new administration. As with almost any organization, the new guys have to flex their muscles; so it was expected that we'd get a lot of feedback each step up the chain. One of the great miracles of all time in the federal government was that our regs were the first ones "unfroze," and in the total rerun of mandatory concurrence and OMB review only **3 WORDS** were changed. This is even more remarkable when one considers that the Department of Energy, which was promulgating the regulations, was an agency the POTUS had declared he was going to close.

When the practitioners of federal mumbo jumbo hear this tale, they typically ask how this could possibly have happened. One possible reason was the interest and the actions of the folks in the field, who were scheduled to be impacted by the revised regs. When the task force got started, we invited a representative of each of the major groups, such as grant recipients, technical people, state energy offices, etc., which would be impacted by the regs, to a meeting and asked them to tell us everything that was wrong with the existing regs. The revisions, therefore, were designed to meet those concerns. We had a strong base of people who wanted those new regs and were truly happy with what we had done. They proved to be a strong political force.

When things go wrong in resolving a water crisis issue,
this method of getting people in the field involved
just might be worth considering.

So that's the way it goes. Sausage and laws—really quite ugly.

PEDDLING INFLUENCE

Much is said about the evils of lobbying. In every election, it becomes a popular whipping boy for the politicians.

Justice Oliver Wendal Holmes is credited with saying, "No generalization is worth a damn, including this one." He was most likely right, but I'd like to claim one exception. *Everyone* has heard a politician decry the evils of lobbying and then heard them proclaim that no lobbyist will ever darken their respective doorsteps. That is a bunch of hooey! I find this cleaned up generalization easy to defend.

I cannot resist making two points here. First, every citizen in this country, in one way or another, pays for a lobbyist. In fact, several lobbyists. The council of churches, the states, the local merchants, etc., etc. all have lobbyists—ALL OF THEM. And they use your tithes, taxes, and fees to pay for an army of lobbyists. And that is not all bad. Ironically, many of the lobbyists for private corporations in D.C. are there to keep the government from doing things that are grossly stupid. All laws seem to have "unintended consequences." Some of those consequences are more blatantly stupid than others.

Our representatives are going to make decisions that impact our daily lives. Those representatives can draw on their infinite internal wisdom and make those decisions in a vacuum, or they can make informed decisions. A good lobbyist will make sure that the politician has the information he or she needs to make the best decision for the constituents they represent.

As noted in Chapter 1, one of my responsibilities at AASA was to find a way to get Congress to help our schools in the 1970s cope with the escalating utility costs. It did not take long to figure out that helping the schools modify their buildings to operate more efficiently would be a very important response to

Figure 9-2. Representative Derek Kilmer in conversation

a critical need. The ultimate result was the Institutional Conservation Program, which many members of AEE cut their teeth on. I was commended by Congress in the Congressional Record for bringing the schools energy plight to its attention. That's lobbying, and I'm proud of it.

Second, at the time I was in the federal government, the lobbyists *representing the federal agencies* by far outnumbered any other group. To the best of my knowledge, that is even more true today. Sometimes they have high-falutin titles, like Congressional Relations Officer, but the effect is the same. If you recall that popular TV series, "The West Wing," the *paid* White House employee, Josh Lyman, spent most of his time persuading or threatening members of Congress. With expert consultants, like Peggy Noonan and Patrick Caddell, the series portrayed how much White House staff time is devoted to influencing congressional members. A standing joke in DC was that if a politician decried lobbying of political action committees, within a week you could expect an invitation to his or her funder. And yes, they almost always showed up on schedule. Next time you see a secretary of some department, or a commissioner testifying before Congress, ask yourself who you think wrote the testimony. Also take a moment to wonder how much time the members of that agency spent visiting with the members of the Congressional committee (or their staff members) before the testimony was drafted. That, folks, is your tax dollars at work.

CUTTING TEETH

In the 70s and 80s, budding energy engineers were cutting teeth in this new field, but they were not alone. We were all on a steep learning curve. What should an audit entail? Should every building be audited? How much would the cost of an audit of a hospital vary from that of an elementary school? What should be the difference in the amount of detail expected and the amount to be charged per square foot? The questions were endless and, in retrospect, the answers were extremely hard to find. Looking back, it is amazing how much of the "traditional audit" embodied in those 1981 regs has stood the test of time and how many aspects of it can still be found in our practices today. The engineers involved in these early exercises were a remarkably savvy group.

A good example of the vagaries of the time can be found in the old literature that refers to much of our "conservation" programs as *insulation* programs. Only after some data were collected and the US DOE put out its first payback charts did people begin to realize that insulation was not the most cost-effective energy conservation opportunity. Those early charts were pretty misleading, but they did reveal that an investment in insulation typically had a lot longer return on investment (ROI) than lighting or controls. Of course, we were dealing with 40 watt fluorescent lamps at the time. We also quickly learned that the best way to save energy was to turn something off, and if you could do it automatically, so much the better.

THE AGENCIES AND CONGRESS

It was not just energy *per se* where the folks inside the agencies needed to get up to speed. From the inside looking out, it was amazing how little the folks in the Forrestal Building (and other agencies for that matter) understood what actually transpired under the dome of the Capital Building. At the request of my boss and my boss' boss, I ended up making presentations at "brown bag" lunches on how Congress works. It came as quite a surprise to those laboring in the vineyards to learn that authorizing legislation only stated the amount that *could* be spent on a program. Then the appropriations committee would recommend how much *should* be allocated, and the budget committee *would actually* set the budget. Even those who put some numbers together to help develop the federal budget each

Figure 9-3. The Capital Building, Washington, D.C.

year did not seem to realize that the president proposes, but it is the congress that disposes. The power of the purse belongs to Congress and all money bills must originate in the House of Representatives.

Even the media has trouble getting it right. The president and his agencies; i.e., the executive branch, can only spend what Congress gives them. We hear regularly of the trillions of dollars the president is spending, but he can only spend what Congress provides.

Learning how our federal government works is a very wide two-way street. Once I had hung up my lobbyist hat and was on the agency payroll, it was made very clear to me that I should not be talking to my friends on The Hill—at least not officially. That dialogue is the prerogative of the Congressional Relations office—a role that is jealously and zealously guarded.

I also learned that the Inspector General's office in any agency has a lot more power than it is perceived to have from the outside. Those lawyers call a lot of the shots.

Another finding that explains a lot is that once the money has been budgeted, the people in an agency or program office regard it as THEIR money. It is not the taxpayers' money. It is not the public's money. It is the government's money—to do with as it sees fit. This also helps explain the "pipeline" argument. Which goes something like: "If it has been budgeted, it's ours. And if we don't spend it this fiscal year, they may think we don't need it and give us less next year." The fiscal year ends on September 30 and the feeding frenzy that goes on that last week of September is something to behold.

Probably the biggest surprise I uncovered was the amount of power the Office of Management and Budget had. I was amazed to find that OMB had the final say on regs, on surveys, on lots of stuff that one would assume was decided in the respective departments. The OMB tentacles have a long reach. Its power is subtle, but it can penetrate deeply into the decision-making process. For instance, I remember vividly the amount of excitement it caused when OMB called to ask if

DOE had an estimate on the number of jobs the ICP created. I was instructed to drop everything and get those numbers (as favorably as possible) to OMB asap.

I walked away from DOE with the impression that OMB is a super-agency with far more influence and power than one gazing from the outside perceives it to have.

Just as New York City has its "social register," DC has the "Green Book." It is published regularly and has a green suede cover. All of the members of Congress and the top agency people are automatically included. A "secret committee" invites other people to be listed in the book. The value of being listed is two-fold: 1) it gets one invited to fill out the list for embassy parties (and you also get to see who else is listed); and 2) it has an incredible section on the appropriate protocol to follow. In the protocol section, I found it interesting to read our copy on the seniority list of departments, which constitutes a "pecking order" of personnel that must be religiously followed. If I were to have some say in developing this list, I would put OMB right up near the top, but then, that wouldn't be very subtle.

INSIDE THE BELTWAY

It is not fair to suggest that the folks at DOE have a monopoly on "fed" thinking. They do not even come close.

An attitude that assumes the government has the answers and seriously needs to take care of all those people out there penetrates all agencies and is pervasive inside the beltway. There is definitely an "inside the beltway" mentality. It is insidious. Without realizing it, living there prompts a growing sense, which is fostered by the local media and the decision-makers, that all things critical happen within the beltway. The beltway, which circumvents D.C., seems to build a fence that establishes the boundary as to all things important. This beltway myopia lends itself into believing it has a certain omnipotence. When Carter became president and brought with him the group affectionately referred to as the "Georgia Mafia," an incident occurred which illustrates this thinking. The Georgia Mafia did not know how to drive in the snow. One day a lady from Atlanta asked if the snow couldn't be outlawed. She was only partly joking.

Inside the beltway is a culture with its own language. It sounds strange when you arrive, but it becomes frighteningly familiar in a very short time. A "cut in funding," for example, means that the planned *increase* in a program or department's budget has been reduced. The next fiscal year will have more money allocated for a project than it has this year—just not as much as hoped for. That's a "cut."

Senator Dirksen has often been quoted for saying, "...A billion here and a billion there, and pretty soon you're spending real money." Today, that *billion* is small change.

Money takes on a different dimension when you are referring to federal dollars. One day, I was doing my lobbying thing and was astonished to hear myself say to other party on the phone, "But it's ONLY 500 million." In my own private world $500,000,000 was (and is) a lot of money. In federal parlance, it is ONLY a pittance.

There are definitely some things a government can do better than anyone else. Unfortunately, a number of people within the government manage to convince themselves that the feds can do everything better than anyone else. This is not limited to one political party. I do not know the ratio today, but at one time I read a government report that said it cost the feds one dollar to spend five cents. Doesn't sound terribly efficient, does it? It certainly lends credence to being "fed up."

THE TIME IS NOW

If we are to become effective stewards of this small spot in the universe we call Earth, we must become incredibly persuasive to a whole range of stakeholders. It has become increasingly apparent, whether we like it or not, our federal government and other governments are going to be involved in some major changes. They must be persuaded and, in turn, become part of the persuasion mechanism. It is essential, therefore, that we know how our government works, how it "thinks," and how it can be persuaded to lead us in the right direction as we tackle some huge problems, such as the coming water crisis.

The inevitability of a water legislation, which will affect the way we live, ending up in the hands of national politicians, staffers on The Hill, agency minions, or the hoi polloi is frightening. There will be a mad scramble to make sure the bill will "make a difference." Not necessarily better—just different.

Chances are that the more our water problems are resolved at the local or regional level, the better off we will be. The window of opportunity to effect those changes is closing rapidly.

Chapter 10

Water 101

If you don't understand water, you won't appreciate it enough or manage it effectively. If you don't know how important it really is, chances are you won't protect it adequately.

We know surprisingly little about water, and the scary part is that we think we know a lot. After all, we've been around water every day of our lives.

We wash in it. We drink it. We swim in it. We are totally familiar with it. But do we know where it came from originally? Do we know it was already here when God created light?

We can run in stark terror from a flash flood and become totally relaxed in a hot tub, but do we know water's unique properties? Or, why it is absolutely vital to life?

Have you ever considered what will happen to our society and to our current lifestyle when we run short of water? We are about to find out.

The signs are all around us. Recently, a lady from California informed me, "I'm from California; I know all about the water shortage."

I still have ridges in my tongue from resisting the temptation to say, "Lady, you don't know the half of it." The problem in California is only the tip of the iceberg. We are in critical trouble, and we don't want to admit it. To put it more bluntly: We are in deep "do-do." We are *stepping in it* and haven't the slightest idea that an insidious problem is creeping up on us from all sides.

Recently, our scientists got all excited when they found evidence of water on Mars. They declared it was very salty water, but water nevertheless. Why the excitement? Because we are about to get an answer to the age-old question: Is there life elsewhere in our universe? Follow the water. If there is life, water will lead us to it.

It's a totally black/white issue. No water. No life.

It's time to look around. We are literally treading water. We have a problem that could destroy us, and yet we go blissfully along like we haven't a care in the world.

The bitter irony is that more than 70 percent of our Earth is covered in water. It's everywhere we look. And the scientists tell us that we have had roughly the same amount of water for billions of years. So what's the problem? As the folks along the banks of the Mississippi can tell you, the problem is that it is not always where we want it when we want it, nor is it always in a form we can use.

We simply can't effectively manage something until we know more about it. Without getting too deep into the weeds, we need to take a look at this thing we call *water*.

A LITTLE HYDROGEN; A LITTLE OXYGEN

Water is a peculiar thing. It is common to all of us. We take it for granted. Most of us remain ignorant of its unique properties, and yet it is one of the most studied liquids of all time. At present, there are about 20 "models" which are used to explain the structure and behavior of water.

Hydrogen is the most common element on Earth. Oxygen is the third most common. It seems logical that they'd rub elbows and figure a way to get together. And they do so by sharing electrons in covalent bonds.

Hydrogen does bond with other elements, but the bond with oxygen is quite unusual. Scientifically, we can say that when two hydrogen get together with one oxygen (H_2O), they create water.

It's the unique way they go about it that matters. A water molecule is a tiny combination of three nuclei and ten electrons, which possess special properties. It is truly unique among more than 15 million chemical species. We have a marvelous love affair between electronegative atoms, hydrogen, and their more positive friends, oxygen. Together they make it possible for the molecules to join in an electrostatic attraction between polar groups called the hydrogen bond.

It is the hydrogen bond that causes water molecules to "stick" together. When the hydrogen and oxygen come together to form H_2O, the bonding electrons are shared unequally. The result is a negative charge at the oxygen end and a slight positive charge from the hydrogen end. This polarity causes the water molecules to be attracted to each other. Ever wonder why ice floats, but other solids sink? Blame it on the bond. Without the bond, our blood would not only boil in novels, but in our veins—a rather uncomfortable thing to contemplate.

The bonding makes all the difference in the world—literally and figuratively.

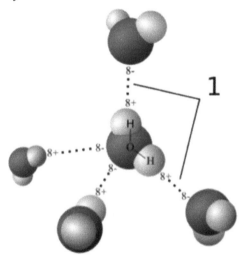

Figure 10-1. Molecular Structure H_2O

Imagine each time you swallow a sip of water, all those hydrogen bonds are hanging onto each other for dear life. But, alas, the attraction is fleeting. All this coming together and breaking apart happens in picoseconds.

With all that proclivity to want to stick together, they hang on a little tighter when there are fewer of them. Thus, water molecules on the surface of water cause surface tension. Done just right, you can actually float a paper clip on water. Surface tension enables the Basilisk lizard to appear to walk on water. It is what makes rain drops round. It is key to water finding its way from the roots of a tree to the leaves. None of it would happen without the hydrogen bond.

It is the hydrogen bond structure that pushes atoms apart as they freeze. Unlike many other substances, water is less dense as a solid, and that's why it floats. If ice didn't float, the *Titanic* might still be chugging across the Atlantic. On the other hand, if ice didn't float, the Arctic and Antarctic would look much different. And those cubes of ice might not rattle so pleasantly in your iced tea.

Of even greater importance to the fish, those in frozen ponds would die each year. Because lakes would effectively freeze from the bottom up instead of top down, the fish would be left without a "warm" part of the lake to live in. Intriguing isn't it that ice actually forms a blanket of insulation for the fish?

Now we come to the matter of making your blood boil, or not. The normal boiling point (also called the atmospheric boiling point) of a liquid is the special case in which the vapor pressure of the liquid equals the defined atmospheric pressure at sea level. At that temperature, the vapor pressure of the liquid becomes sufficient to overcome atmospheric pressure and allows bubbles of vapor to form inside the bulk of the liquid. Because those water molecules like to cling together, it takes more

Figure 10-2. The Boiling Point

vapor pressure for the liquid to overcome the atmospheric pressure. Once more the hydrogen bond triumphs.

One more factoid, which will probably be of little use to you as you tackle water management issues, but interesting nonetheless: Water is most dense at 4°C, so that is always the temperature at the bottom of the ocean. Below 4°, water is starting to form crystals and expand. Above 4°, it is less dense and will rise. Can't you see yourself dropping this tantalizing information at your next social gathering?

A QUESTION FOR THE AGES

So where did all our water come from? We don't know for sure. The Bible tells us that there was water here when God created light, saying, "The earth was without form and void, and darkness was upon the face of the deep; and the Spirit of God was moving over the face of the waters. And God said, 'Let there be light'."

It is believed that, at some point in time, all water came from space and was probably in exactly the form it is in now. It is said that water is 4.3-4.5 billion years old. Surprisingly, most water on Earth is not in clouds, lakes, rivers, oceans or aquifers, nor is it in the form of ice, liquid or vapor. There exists a fourth form of water locked in the Earth's mantle about 410 kilometers (255 miles) down. It is not mixed in, but baked into the molecular structure of the stone itself. There is an estimated "5 oceans" of water captured in the Earth's mantle. Since the mantle is deep in our Earth, no one knows for sure how much water exists at that level. We do know, however, that there is typically one hydrogen (H) atom wandering around on its own while another pairs up with oxygen (OH). When rock, such as serpentine, is treated, they all get together and come out as water. Some believe that much of the water on Earth's surface could have come from this imbedded water through volcanic eruptions.

All that water came from somewhere and covers 71 percent of the Earth's surface. Of all life that inhabits Earth, about 90 percent lives in water. Perhaps this gives credence to the thinking by some that homo sapiens evolved from marine life. Or, such thinking might be supported by the fact that about 55-60 percent of our bodies are water, and water is absolutely essential to life.

Ironically, water is critical to our very being,
but it is incredibly cheap.

Mother Nature has set up a marvelous cleansing system through an evaporation and precipitation cycle. Quantities are so huge they are measured in units called cubic kilometers. A single cubic kilometer holds 250 billion gallons. (It would cover Manhattan 37 feet deep.) Every hour, 50 cubic kilometers (15 trillion gallons) of water evaporate from our oceans. That's worth repeating: *one hour ... 15 trillion gallons.* To help put that in perspective, the US uses about 410 billion gallons a day. Or, consider that an inch of rain on half an acre is roughly 13,577 gallons. No matter how we look at it, a cubic kilometer of water is a humongous amount.

Another bunch of water we should be aware of is called *biological water.* It's worth noting because it amounts to about 1,120 cubic kilometers, or about one tenth of the water cycling through the atmosphere. This water is temporarily locked within everything alive, including us. Humans account for about 35 billion gallons of walking-around water.

For those who want to look farther into how much water is on Earth, please see *Water in Crisis* by Igor Shiklomanov, edited in 1993 by Peter Gleick.[10-1]

THE FAUCET

We all know where water comes from: It comes out of the faucet. Just a twist of the faucet, and we expect clean, potable water to emerge. In reporting on the lead in drinking water concerns, Shepherd Smith stated, "One of the things you deserve from your government is clean water."[10-2] For the young, it's always been there, but I can remember my grandmother going "out back" to pump water by hand. I also remember my grandparents' delight when they got indoor plumbing.

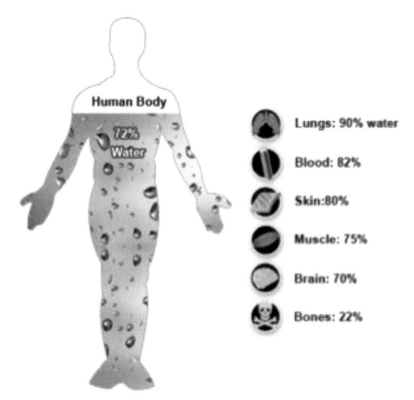

Figure 10-3. We Are Water

For most of us, a twist of the faucet and, ta-da, clean water is there. But not for everyone. Various reports have indicated that water pollution is the leading worldwide cause of death and disease, and that more than 14,000 deaths *per day* can be attributed to the consumption of contaminated water. In India an estimated 580 people die of water pollution related illness every day. In 2007, an estimated half a billion Chinese had no access to safe drinking water, and as much as 90 percent of the water in the cities of China is polluted. In the most recent national report on water quality in the United States, 45 percent of assessed stream miles, 47% of assessed lake acres, and 32 percent of square miles of assessed bays and estuaries were classified as polluted. Unlike many countries, we do not need to rely on this water directly. When it is used in a community, it is treated and filtered first.

Figure 10-4. Dying for a Drink of Water

Our concerns are not limited to what comes out of the tap. Nearly 1 in 8 (12.1 percent) of routine pool investigations in 2008 identified serious violations which threatened health and safety, leading to immediate pool closures. Among the villains was Cryptosporidium (Crypto), which has proven to be an extremely chlorine-tolerant parasite.

SHADES OF GREY

When it comes to water quality, we have our own shades of grey. We have the clear potable water we drink, we have greywater, and we have blackwater. Without getting into these classifications too deep, we still need to understand their differences if we are going to manage them effectively.

The US is fortunate to have some of the safest drinking water in the world. In the US, the quality of water is regulated by the Environmental Protection Agency (EPA). The EPA sets standards for testing water. The agency also provides standards for more than 90 different contaminants that can be found in water, including E.coli, Salmonella, and Crypto. The agency has a recommended "action level" for lead but no standard. The agency also sets standards for disinfection byproducts. More information can be found at EPA's site, Drinking Water Contaminants.

Every community water supplier must provide its consumers with an annual report, *Consumer Confidence Report*. The report must include information on the water's source(s) and the levels of contaminants found in the water.

Greywater. In the literature, you can find *greywater* as two words, hyphenated, or spelled graywater. It's also called sullage. It's all the same water. Basically, it is wastewater from households and office buildings. It is the wastewater from sinks, baths, showers, clothes washing machines, dish washing machines, etc. By definition, it is not supposed to have any pathogens or fecal matter. To put it more bluntly, it is *all* wastewater except for what you flush down the toilet.

It is aptly named *wastewater* because we are wasting it. It is perfectly good water that can be reused for other purposes. Greywater is generally safer to handle than blackwater and easier to treat. Greywater, therefore, can be cleaned to use as drinking water, but is more typically reused onsite for landscape work, crop irrigation, toilet flushing and other non-potable needs. Because it is possible that some pathogens may be in

greywater, it is recommended that the water for irrigation purposes be applied directly and not through a sprinkler system.
Such a system can create airborne contaminants. In the US,
most states have adopted the *International Plumbing Code* or
the *Uniform Plumbing Code* to guide subsurface irrigation and
other reuse applications. State laws and permit procedures are
changing rapidly, so it is important that those who wish to reuse
water check locally for guidance.

Our concern in the US is drought areas or areas served by
the Ogallala aquifer as shown in the darkened area on the map
in Figure 10-5. The aquifer covers a lot of territory, and many
people depend on water from it. The aquifer is drying up. Some
estimates give it 20 years. That means we should have been
planning *yesterday* to more judiciously use the precious amount
that is still there. If I lived there, I would be pounding on a lot
of desks, asking what we are doing about it.

Figure 10-5. The Ogallala Aquifer

Can we keep drawing water out of aquifers as if there were no tomorrow? Everyday we water the lawn with water our future children will truly need for drinking.

On the good news front, the United States Department of Agriculture has just awarded a $10 million grant to address the Ogallala problem. A broad group is involved in the project, including Colorado State University, University of Nebraska, Oklahoma State University, New Mexico State University, Texas Tech University, West Texas A&M University, the Texas A&M AgriLife Research/Extension Services and the USDA Agricultural Research Service. There is great hope for an idea exchange among other actions. Dr. Brent Auvermann from AgriLife observed, "For many years, researchers have been developing more efficient irrigation techniques, water-efficient crop varieties and water-conserving soil management methods. But we are still draining the aquifer. It's time to put all of the pieces together, and it's time for each of the affected states to learn from the others." The group's first meeting was in March of 2016.[10-3]

BLACKWATER

When steps are taken to keep greywater apart from toilet water, it is called source separation. When mixed together with toilet wastewater, it is called sewage or blackwater. Then all of it must be treated as sewage in treatment plants, or onsite sewage facilities, such as septic systems. When it is kept separate, the reuse options are greater and the treatment less demanding. Source separation is one commonly applied principle in ecological sanitation approaches.

Blackwater is a term used to describe wastewater which contains feces, urine, and flush water from toilets along with cleansing water for these fixtures. Increasingly, facilities are plumbed to separate greywater and blackwater.

Because blackwater contains pathogens that must be decomposted before they can be released safely into the environ-ment, it is heated typically using thermophilic microorganisms. One year after treatment, it is considered safe to use as fertilizer.

The amount of blackwater being generated into the system can be reduced by the use of composting toilets and vermicom-posting toilets. Proponents note that this procedure will allow the reuse of nutrients found therein. As unbelievable as it seems, this may be part of our future.

TREATMENT

Nearly all water needs to be treated to meet the public health standards before dispensing it to a community. The most common treatment for a community's water system involves coagulation and flocculation. A positive charge is added to the water (often by chemicals), which neutralizes the negative charges of dirt and other dissolved particles. The particles then bind together and the weight of these larger particles causes them to sink, creating sedimentation.

Then, the cleaner water passes through filters of varying compositions (sand, gravel and charcoal). Finally a disinfectant, such as chlorine, may be added to kill any remaining parasites, bacteria, and viruses and to protect the water from germs.

Treatment of the water will vary with its incoming quality. Because it's typically exposed to nature, surface water is apt to require more treatment than ground water.

To prevent tooth decay, community water is typically treat-ed with fluoride. This is a prescribed procedure and has been singled out as one of the 10 great public health achievements of the 20th century.

There is nothing terribly mysterious about treating water. A wastewater treatment site is common in most communities. It gets more interesting if you are in a floating hotel, which is what a cruise ship is. In *Sustainability Management Handbook*[10-4], I shared a case study of the great work Celebrity Cruise Lines is doing. Unlike some other lines, which have made some unhappy headlines since that book was published, Celebrity ships always have the needed water. They have their own desalination plants. They also pay meticulous attention to wastewater; and, in fact, they treat their blackwater to meet or exceed public health standards, before it is discharged into the ocean.

One is left to wonder: If we can do all that in a floating hotel, can't we do a better job for the villages in Africa?

WATER AS A SOLVENT

Water is often referred to as a universal solvent. It dissolves almost everything. Water does this by surrounding charged particles and in effect "pulling' them into a solution. When you put your hands under the faucet, you expect to come away with clean hands. It's a miracle we experience every day.

Any substance with a negative electric charge can dissolve in water. This also explains why something like oil, which is a nonpolar molecule, will not dissolve in water.

Water is not, however, at all consistent in its dissolving behavior. When you put regular table salt (NACL) in water, the negative oxygen ends of a water molecule surround the positive sodium ions (NA+) while the positive hydrogen ends surround the negative chlorine ions (CL-). I doubt, however, that you can taste the difference when you sip some salted water.

Water's ability to dissolve most everything is what makes it such an effective cleaning agent. Unfortunately, this causes lead in drinking water and headaches for Flint, Michigan, city administrators.

The Flint problem, however, was not a water issue as much as it was a management issue, or maybe an intelligence issue.

When you test the water and find
harmful contaminants in the water,
you let people know.

For the EPA to tell the city officials and then sit on the sidelines for months while vital information was not provided the public seems like gross negligence. EPA claims they were operating within the regulations. But where in the regulations does it say you should sit by and let children drink water harmful to their health? It appears the city officials had decided to save money by not buying Detroit water but to get it locally. When they discovered they had made a mistake, they did not own up. Flint, the state, and the EPA offer us a lesson in all the circular finger pointing which is now going on. They are looking for a fall guy. If a similar situation ends up in your backyard, it could be you. It's all about communications.

The problem with such mismanagement is that it makes us all feel vulnerable. Listen to the hue and cry to get local water tested and to inform the public. It also makes us more conscious of our dependence on water (and our dependence on government), and gives us a small example of the multiple public outcries we *will get* as the availability of clean water becomes an issue in our society.

The Flint problem has generated a lot of interest in lead in drinking water. This is not a new problem. In *The Great Lead Water Pipe Disaster*,[10-5] Werner Troesken has given us an excellent review of the history of lead in water. He points out the associated sickness, premature death, political inaction, and social denial related to lead pipes in local water systems. The tragic thing is that doctors and scientists started noting cases of acute illness and death that could be attributed to lead in public water as far back as the middle of the 19th century. An editorial in *The New York Herald* called on the city to study the matter at that time. No action was reportedly taken until 1992! Troesken offers old and new science of lead exposure, including an explanation of why soft water suffers more harmful effects than hard water. In addition to the books listed in the references, more information can be found at http://marginalrevolution.com/#sthash. ICSOU7aS.dpuf.

Flint, however, opened the door. Expect a lot of water

testing and alarm about lead in the drinking water in the days ahead. Expect the EPA to be the one with the megaphone. Chances are there will be new regulations, a bigger water group within EPA, and more GS upgrades and pay raises. That translates to more money and more government intervention.

CLEAN WATER

There are many stages of "clean." The water used in some high-tech manufacturing takes it to a new level.

The ultimate is the ultra-pure water (UPW) used to produce microchips. UPW is purified to uncommonly stringent specifications and becomes a supersolvent. The literature describes anywhere from three to eighteen steps of cleaning UPW; then, it becomes hungry for anything. In fact it becomes so thoroughly cleansed that it will actually function like poison and eat tissue of the human body.

Multiple standards have been developed and published associated with the production of UPW. The most recognized requirements for UPW quality are documented in the American Society for Testing and Materials International (ASIM) in D5127, "Standard Guide for Ultra-Pure Water Used in the Electronics and Semiconductor Industries," and the "Semiconductor Equipment and Materials International (SEMI) F63, Guide" for ultrapure water used in semiconductor processing.

UPW seems like a tangent to our water planning, but it is critical to a slice of industry. Look it up on Wikipedia and you will find 25 pages on the topic. It's an excellent example of a special needs industry that just might be in your community.

The reason for all the attention to *clean* water is blatantly simple. Value-added technologies and products command better prices than commodities. If you can add that value for less than the price it commands, you have the makings of a business. Dow Energy & Water Solutions (DE&WS) is considered to have the competitive advantage in this field. Larry Ryan, the business president of DE&WS, reports that the company spends more than $1.6 billion each year on R&D, and adds, "We will continue to invest heavily, because we see that as an advantage we bring to the market place."

As discussed in the next chapter, water is indispensable to the manufacturing of many products. A drop in water supply can have devastating effects on our manufacturing capabilities and our standard of living. That is why *Water Poverty* is such an apt title for this book. As we vie for water in our daily lives, its absence has the potential to cut to the very marrow of our bones.

One of the great ironies of life is that we essentially ignore something we can't live without. We are living in a state of denial. That, however, is about to come to a screeching halt.

MYSTERIOUS WATER

No discussion of water would be complete without a nod to the mysterious art of water witching, or dousing.

When we were young and foolish, my husband Jim and I decided to move back to the family farm at Randle, Washington. Since the old farmhouse had burned down and the valley was prone to flooding, we decided to build on higher ground on the other side of Syler Creek. All went well until we tried to find water. After pouring money into the ground for three dry holes, a good friend and respected member of the community, our state representative, Harry Syler, came over and rather sheepishly offered to try water witching. Harry proclaimed he did not really believe in water witching, but admitted with a shrug of his shoulders that it sometimes worked.

He cut a "Y" shaped branch from a nearby tree and pro-ceeded to wander around. When the branch turned abruptly down, he said we would find water below where he was standing. When he saw the skeptical look on my face, he put the branch in my hands and put his hands over mine. We then proceeded to cross the spot he'd identified. The branch twisted downward in my hands so hard it tore the bark off the limb.

Some explain this "witchery" as ideomotor and attribute the action to a subconscious reflexive response. It's difficult to see how that would explain the action in MY hands.

A couple of days later, the guy who was going to do the drilling showed up. He took an old rusty wire out of the back of his battered pick-up and held it over the spot. After watching the wire bounce around a bit, he declared we'd find water at 57 feet.

He was wrong!

We found an abundant supply of water at 55 feet. To this day, no one has adequately explained to me how, or why, that all happened.

You can't help but love water when it responds like that.

SAY GOODBYE

The Golden Age of Water is over. For some, in remote areas of Africa, it has been over for a long time (if it was ever there to begin with). For most, we have forgotten how our grandmothers had to pump water by hand in the good old days. For most, outhouses are something out of a comic strip. Few remember how ecstatic our grandparents were when they suddenly had indoor plumbing. Now, if "House Hunters' is to be believed, we cannot understand how a family could possibly survive with only one bathroom.

We have been spoiled. Tragically, we waste water all the time. With few exceptions, we do an incredibly poor job of managing water. That's what happens when it is abundant and cheap. For those of us with a few grey hairs, it sounds a lot like those days when we had $.26/gallon gas and not a worry in the world.

The Golden Age of Water, however, is coming to a close. We are going to have to wise up ... fast.

In summary, water has been described as elegant, smart, sly and charming. Part of its charm can be attributed to its ability to be simple and complex at the same time. When it comes to simple, it's only three atoms—two hydrogen and one oxygen. By comparison, a simple molecule of DNA has 204 billion atoms.

That simple water molecule has some incredible properties. Due to a slight polarity (one side positive and the other negative) it has a complicated set of properties. Consider that it is literally ubiquitous while it is transparent, reflects light, absorbs heat, cleans, cushions a rocket blast and yet is hard enough to really hurt when you do a belly flop into a pool.

Our emotional attachment to water is puzzling. We find it soothing. A walk on the beach is the ultimate indulgence for some. We find waterfalls awesome. Think of the billions spent on boats just so we can putter around on water. In preparation for this book, I asked many, many people what they found soothing about looking at water and not one of them could come up with an answer. They seemed surprised that they should even be asked. It just was.

Chapter 11

Troubled Waters

Before we try to build any bridges over our troubled waters, we need to get a grip on the many demands placed for water. Who uses it? How do they use it? Who needs it most?

This chapter offers a rich sampling of how many ways we count on water to meet our needs. No matter which way we turn, water is a critical part of our world. It is essential to our economy, necessary to our environment and absolutely vital to life itself. We cannot sustain our way of life or our world without water. A serious decline in the amount of water available in any locality will drastically change the way we do business and the way we live.

Whether we talk about the *need* for water, or the *demand* for water, it is troubling that we aren't really aware of the vital role it plays in our lives. It is even more troubling that we are facing a water crisis and that so few seem to be planning for it.

Before we consider specific demands for water, let me pose a problem that will bring home the emotional and political overtones water scarcity carries with it: If a little girl needs water to live, and the neighbor has water but is not willing to share, what do you think that little girl's dad will do? If it is a region or a nation that desperately needs water where lives and jobs depend on it, what do you think people will do?

Water Poverty is on the horizon, and it is not pretty.

A STREAM OF THOUGHT

Water is the most important substance in our lives. Whether we are contemplating clouds on the horizon or a puddle in the road, it is ubiquitous. We expect water vapor in the sky, rain and puddles. Water dictates our weather and shapes our geography. The Grand Canyon is a marvelous example of water at work.

In a strange way the invisibility of our current water systems is a tribute to our technological progress in the industrialized nations. Unfortunately, that very invisibility leads to a sense of false confidence. When we are done with water, it nicely disappears down the drain, and we forget about it. Any conceivable water problems seem to swirl down the drain just as fast.

There was a day when we seemed more aware of water. Notice how many cities are founded near bodies of water. Even now, water offers a major transportation pathway. A search of ancient history reveals that the Roman aqueducts were actually treated as monuments to engineers and to water. In fact, a study of the problems faced and the solutions found in constructing those aqueducts would make any engineer proud. It's worth a little detour to read about the incredible innovations, such as the siphons used to send water uphill. Rome had huge water problems. The engineers came through. I suspect we'll do it again.

Before we get into specific demands on water, it seems appropriate to pause and recap its virtues.

APPRECIATION TIME

Water is a miracle.

Another liquid of roughly the same mass might boil at 100° Fahrenheit, but water doesn't. Why? Because it makes up more than 50 percent of our bodies, it would be mighty uncomfortable on a hot day if it did boil at 100° Fahrenheit.

Any self-respecting solid would sink in its own liquid when given the opportunity. But water doesn't. The crazy stuff floats instead.

Given the right conditions, it turns into beautiful snow-flakes. I'm told no two are alike. Since I haven't checked all the snowflakes out there, I'll have to take their word for it. But how does it do that? When we were in grade school, we all picked up our scissors and attempted to cut the intricate design of a snowflake. Remember folding the paper and trying to cut through several layers? Did you ever wonder at the time how snowflakes form and why they are each different? It's time we really appreciate the uniqueness of water.

It all comes down to that hydrogen bond we discussed in the last chapter. It's what gives water its unusual properties and makes it essential to so many functions in our life.

As already noted, we have had water for at least 4 billion years. During that time, we have always had a drought somewhere, sometime. *Always*. When a drought occurs where people are, we notice. When it occurs where a lot of people are, like California, we really notice. And when it lasts a long time, it makes headlines.

Our problem is that water is not always *there* where we want it, when we need it, or in the form or quality we must have. We only have to see bottled water being flown in to flooded areas to realize how complex the problem is.

It is hard to rationalize how saving water in the United States will save the children in Africa. The problem is local. The solutions are local. We, however, will not be in a position to help those struggling to find clean water if we allow ourselves to drift toward that level of water poverty. As we look around, it is past time to get our house in order. A first step might be to start regarding our drinking water as the precious commodity it is.

THE DEMAND FOR WATER

As we consider the many demands placed on water, it is only appropriate that we start with our old favorite, energy, and the almost symbiotic relationship that exists between energy and water.

ENERGY AND WATER

Water and energy are mutually dependent, and nearly in-
separable. Energy places the greatest demands on water use in
our country. Forth-nine percent of our domestic consumption
of water goes to the electric utilities, and that does not include
hydropower. That's about one and a half times the water used
by all of our farmers. To break it down, a day of electricity in
our lives requires about 250 gallons of water per person, which
is more than the average person uses each day.

A related factor has mostly escaped us: Water absorbs energy. Most cities around the world get their water far upstream,
then use the water, pollute it, clean it, and put it back in the
river far downstream. This process, described as a linear para-
digm, uses a lot of energy. The desalination plants in Australia
and San Diego use lots of energy. Those who care about using
our limited energy resources wisely should consider the impact
water treatments are, and will have, on those resources in the
days and years ahead.

Recently, the ACEEE and the Alliance for Water Efficiency
published a white paper that described the co-dependence of
water and energy. They highlighted the following facts:

• Sourcing, moving, treating, heating, collecting, re-treating,
  and disposing of water in 2005 was consuming 19 percent
  of California's electricity, 30 percent of its natural gas, and
  88 billion gallons of diesel fuel annually.

• The River Network found that energy use in 2009 for water
  services accounts for 13 percent of US electricity consumption—at least 520 million megawatt-hours annually.

• Thermoelectric power accounted for an estimated 49 percent of US water withdrawals and 53 percent of fresh sur-
  face-water withdrawals in 2005.[11-1]

Water is used in making other sources of energy as well. It
takes 17 gallons of water to make a gallon of ethanol. As frack-

ing is making gas more readily available, it's time to pause and weigh which is more valuable: the ethanol or the water. As we weigh the relative benefits of ethanol and water, we need more information on the relative environmental impact of natural gas and ethanol. This highlights how critical it is to recognize that water decisions are not, and cannot be, made in isolation. Every major water decision will have far-reaching economic and environmental variables to consider.

Of course, there is also the question of how much water fracking takes and whether potable water, wastewater or recycled water is being used. Hydraulic fracking is regarded as one of the key methods of extracting unconventional oil and gas resources. The International Energy Agency has estimated that the recoverable resources of shale gas by this method are about 208 trillion cubic meters. The literature on fracking also indicates that a typical fracking operation uses about 40,000 gallons of chemicals.

The "hydraulic" part means liquid pressure is used. It is acknowledged that about 90 percent of that liquid is water. Whether it is our insensitivity to the looming water crisis or the fact that no one wants to say, but...

it is inordinately hard to find any figures
on how much water fracking is using.

One source, which was not fully credited, estimated that it took 5 million gallons of water to serve one operation over its lifetime. Any details about the disposal or reuse of the water are also sparse in the literature, but Figure 11-1 from the EPA suggests that some reuse is part of the procedure.

It's increasingly apparent that we can't make energy decisions without considering the water implications, and equally important that our water decisions take into account the impact on our energy. It takes water to provide energy, and it's going to take more and more energy to provide the water we need. It is a double-edged sword with no easy answer.

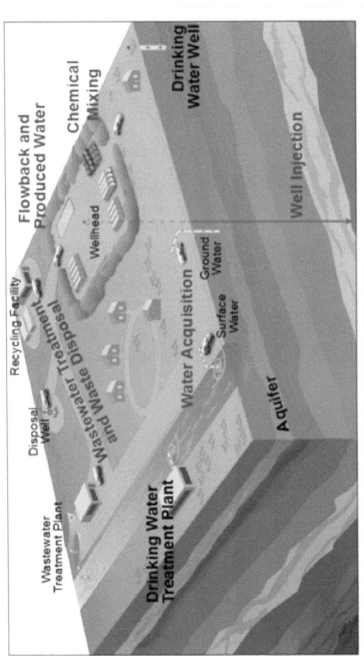

Figure 11-1. The Hydraulic Fracturing Water Cycle

Considerable concern has been raised regarding the effect of fracking on ground water. Since most fracking activity is far below ground water level, further definitive research on any relationship is needed.

FRIEND AND FOE

Whether it is trying to get rid of the melting snow in Boston, or another headline about the huge devastation the Mississippi has caused by flooding its banks, or the serious drought in California, we have a perennial problem of too much or too little water. But these concerns are temporary. The villages in Africa where children are dying because they do not have clean water to drink offer a far more critical and lasting concern.

In Asia, groundwater, reservoirs, and even the surrounding sea water still suffer from the contamination from the crippled Japanese nuclear power plant in Fukushima.

THE NEVER ENDING NEED

Estimations on the percent of our bodies that is water vary, but conservatively water represents about 60 percent in men and 55 percent in women. Our blood is 83 percent water. While you are reading this, you blink your eyes to be sure your eyeballs stay moist.

We are walking around testimony to how essential water is. Figure 11-2 identifies many of the ways water is used in our body. It only highlights, however, how indispensable it is to our well-being. It is impossible to overstate the crucial role it plays in our lives. Maybe the best way to stress this is to make the simple observation that without water we wouldn't be here.

Water has a critical role to play in our economy and in industrial processes as well. Ninety-one gallons are needed to make a pair of jeans. Well, those old raggedy ones are fashionable, so we'll be okay for a while. But what about the factory

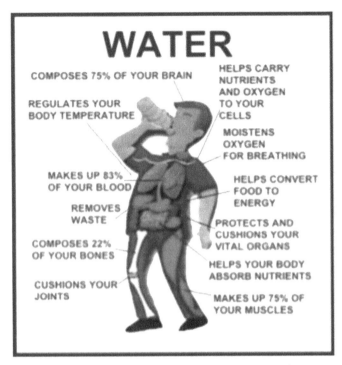

Figure 11-2. Water and the Human Body

workers that make new jeans. Lack of water will cost jobs in many lines of work.

Just think about the water required to make soda. It takes 5 liters of water to fill a 2-liter bottle of Coca Cola. Hidden consumption is everywhere.

One place where water loss deserves special mention is *leaks*. According to a recent report from the Urban Land Institute and Ernst & Young, our greatest challenge in the water world is our deteriorating water system, which loses 16 percent each day from leaks in water pipelines. Our dilapidated infrastructure is leaking about a day's worth of water every six days or about 1.25 trillion gallons every year. There is little comfort in knowing that the UK (19 percent) and France (26 percent) lose more. With all the talk about infrastructure and shovel-ready jobs, we ought to be able to do something about leakage at that rate.

Those leaky old pipes are losing more water each year than the annual amount used by the residents of Los Angeles, Miami and Chicago—*combined*. Sadly, low water bills and the lack of funding for improvements stand in the way of getting this critical need taken care of.

We tend to think of water in terms of our immediate contact with it. But the energy numbers above reveal how imbedded water is in everything around us. Water scarcity will have a tremendous domino effect on things we typically don't associate with water. We are only beginning to recognize the priorities we will need to set and the decisions that will need to be made.

If we follow our historical pattern of fervent denial until the crisis hits, the chances of making quality decisions amid the panic and chaos are not good. A classic example is the response in Barcelona where city officials commissioned both the *Sichem Defender* and eventually one other ship to bring in enough water for 5 million people. Relief from the first shipload lasted 32 minutes. This noble effort to show the city was trying to manage its crisis cost of $30 million and left behind the *Sichem Defender* as a symbol of poor planning.

Orme, Tennessee, became the poster child of water scarcity in 2007 when the spring went dry. Earlier, the city had passed up laying a pipeline and now had to rely on the fire truck running to nearby Bridgeport 10 times a day for its water. For months, the mayor opened the valve at 6:00 p.m. to let the water flow for precisely three hours. During that time the residents could wash dishes and clothes, cook, and fill some buckets to hopefully see them through the next 21 hours. Not surprisingly, the pipeline was built about 4 months later. Orme is not a quaint story to add color to the narrative; it offers a lesson.

GOING FROM NEED TO DEMAND

Setting priorities will not be easy. We will be faced with obvious *needs*; but **demand** from the vociferous few is apt to drown out the voices of need.

Figure 11-3. Have Some

Imagine a world with a lot fewer cell phones. What if water scarcity occurs where they make ultra-pure water (UPW), but there is a great need for that water stock for the community's health? A cut-back on UPW would put a huge dent in electronics manufacturing. The ripple effect—no, make that the *tsunami* effect—would be felt far and wide. The world would suddenly have fewer smart phones, tablets, and eventually computers. The economic impact would be gigantic. The DEMAND would be huge. How could we possibly live without our cell phones? Our computers? Oh, and what about the *people* who need clean water in that manufacturing town?

Water's ability to act as a solvent can also pose problems. One of the big ones is lead in drinking water. The problems in Flint may only be a blip on our radar. Most lead in drinking water comes from the plumbing. It is especially bad in old plumbing where the water has been allowed to sit for a time, such as schools during vacations. Neither the EPA nor the CDC has stated a concentration level that would constitute a "do not drink" order. In fact, as recently as 2004, the EPA pulled a statement about 40 parts per billion (ppb) when the levels of lead in the Washington, DC, water got a lot of publicity.

EPA regulations do require that customers be advised if the level exceeds 15 ppb. This is referred to as an "action level." A recent *USA Today* article reported that some 350 water systems (10 percent) in the US show lead concentrations above 15 ppb.[11-2] Because the concentrations are a result of both the concentration at the source and plumbing problems, levels vary from house to house. They have found levels in Flint as high as 12,000 ppb. The EPA has described any level above 5,000 ppb as "hazardous waste."

The properties of water lead to some unusual applications. One of the most striking is its use to absorb the sound shocks in firing spacecraft. Without the sound cushion that water provides, the sound reverberations would tear apart a spacecraft. The problem is that it uses an incredible amount of water (about 300,000 gallons) every time we have "blast off." Where in our water priorities do we put our desire, if not demand, for space travel?

The list of water crises which threaten us is long and getting longer. The US alone has great needs. There does not seem to be an end in sight. Our growing US population (expected to add 120 million additional residents in the next 40 years) will put even more demand on water-constrained parts of our country.

Underlying and underscoring this growing dilemma is the inability of people to recognize the threat. This is not an *impending* threat. This is a *here and now* threat. It is slowly changing where we live and how we live. Denial will not make it go away.

The next time you turn on the faucet to get a glass of water from the tap, enjoy its fresh cool taste. Then, think about the people in China, India, Mexico, Brazil, or South Africa who are aware that it is not always advisable to drink the water. The quality is uncertain, and they might be putting their health in jeopardy.

CASE STUDY: AUSTRALIA

A 15-year drought in Australia prompted huge investments in reducing water use and developing new water sources. It has changed the way people live in Australia. Recent
statistics show that national consumption has been brought
down 20 percent. Water recycling was targeted to hit 30 percent
by 2015, and six desalination plants are scheduled for start-up
this year. Though the drought has lessened, the scare has left
a lasting effect.

After the Australian drought eased in 2012, a study of
lessons learned determined the importance of lessons yet to be
grasped, including:

a)   do not underestimate the value of efficiency and conserva-
     tion;

b)   effective communications is essential;

c)   there is a tendency to focus on residential problems and
     overlook businesses;

d)   alternate supplies through desalination and recycling are
     needed (though rate payers are reluctant to cover the costs
     once the drought eases);

e)   price adjustments must consider the water utilities' lost
     revenue and the customers' ability to pay; and

f)   greater emphasis on climate risk management and pre-
     paredness is needed.

They now have plans ready to start building desalination
and recycling plants if the water depletion hits a certain trigger level. This preparedness would have the plants operational
by the time the traditional source availability hits bottom. The
reader is encouraged to review the extensive literature available
on the Australian experience.

CASE STUDY: CALIFORNIA

No discussion of water in the second decade of the 21st century would be complete without a hard look at the California drought.

Figure 11-4 shows an irrigation ditch in Richvale, California. This is an all too familiar site in recent years in California farmlands. We heard about the residents having problems, but the losses in the agricultural industry have been huge. The subsequent effect on the price of food could be felt throughout the West and beyond. It was a strong reminder that drought effects may seem local, but their ripple reaches far.

Having survived earthquakes, energy crises with rolling blackouts and a budget collapse, California has in recent years seen a return to a thriving economy, population growth, and even a surplus in the state budget. But the specter of persistent drought has raised questions about California's future.

More than 38 million people live in California today—more than twice the number living there in 1960. The economy is

Figure 11-4. Irrigation Ditch in California

now ranked 7th largest in the world at USD 2.2 trillion. What happens in California has far-reaching repercussions.

Despite some good public relations of lush Palm Springs and other California oases, the state has a long history of dry spells. Figure 11-5 is an April 2016 map of California from the United States Department of Agriculture depicting how wide-spread the current drought is. The darkest area is described as severe drought and the lighter shades depict the drought impact to varying degrees.

Even though earlier droughts brought about some modest changes in water use, the current drought seemed to catch officials unprepared to manage its impact.

While consensus places the beginning of the current California drought in 2011, Governor Brown did not order his proclamation until April, 2014. The proclamation then embraced

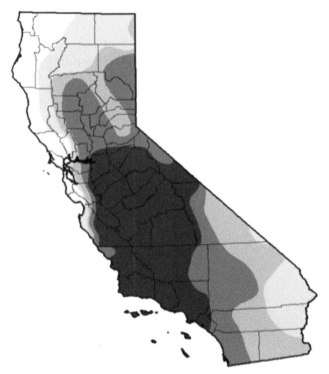

Figure 11-5. Map of California's Drought

what some called draconian measures. Appendix B has a copy of the proclamation. In fairness to the governor, it should be noted that some cling to the idea that one good rain (or even one wet winter) will fix the problem.

In fact in May of 2014, Katherine Mieszkowski wrote in *The Center for Investigative Reporting* that the drought laws among California's biggest water users was largely being ignored. According to Mieszkowski, there are few consequences if the water districts do not comply with the mandates.

The governor ordered a 25 percent reduction in residential water use. Cities, such as Palm Springs, often described as "America's desert oasis," ordered 50 percent cuts in water use in the city agencies and now have plans underway to replace the lush green lawns and lovely annual flower beds with native landscapes. That means rocks and desert plants.

The people in California, long used to a daily plunge in the pool, are becoming accustomed to sites like the one shown in Figure 11-6.

The 25 percent cut did not apply to farmers, but most California watchers think the farms are next on the cutting board. The implications for the agricultural economy in California are frightening. The cost of food will escalate.

Efforts to grapple with the problem have been uneven and too frequently ignored. Those who are trying to comply are not pleased with the way others ignore the situation. Figure 11-7 depicts the contrast some see and feel.

Fortunately, many leaders in California are now laying the groundwork for what they consider the new normal. Dr. Kevin Starr, a state historian at the University of Southern California, has observed in response to some doomsayers that the state is not going to go under, "...but we are going to have to go in a different way."

The Mayor of Los Angeles, Eric Garcetti, has pointed to dry, truly desert communities, such as Phoenix and Las Vegas, and commented that they "...have shown the ability to have economic growth."

Figure 11-6. Swimmers Must Shower Before Entering Pool

Figure 11-7. Green is More Fun

In response to the claims that the drought would prompt industries to move away or stop adding jobs, Allan Zaremberg, president of the California Chamber of Commerce, rejected the idea, observing, "The rest of the economy is managing it, learning how to deal with it."

One of the best takeaways from the California experience is the importance of leaders to take such positive positions and calm the growing hysteria.

A final California observation seems appropriate. While the total water on Earth stays about the same, droughts exist somewhere at all times. When they persist in areas of dense population, they garner the attention that California has.

Sometimes a bit of whimsy helps lighten the load. Figure 11-8 offers a good note on which to close the California case study.

THE "YUCK" FACTOR

That "closed loop" phrase is easy to slip by, but it represents a huge change in how we use water and how we live. A closed loop means the water is reused. What you are drinking could be someone else's dishwater, or ….

It helps get the "yuck" factor out of our thinking when we realize that we have the same water on Earth that we had over 4 billion years ago, and it has been used and reused over and over again. Our planet and its weather system does a great job of purifying water. We are already drinking reused water. Get used to it.

It is time that we woke up to the fact that we should not be using potable water to keep our yard green. We currently water our daisies with water that humans in other parts of the world desperately need to stay alive. Granted, saving the water here may not save those people, but we continue to ignore how precious our water is. The water we discard from processing plants could serve a village.

Figure 11-8. Drought and Diving?

In an age of water poverty,
we must learn to reuse water.

It is incredibly tempting to go on and on—to offer a really long list as examples, because there are so many applications of water everywhere. But you get the idea. It's important, however, to take a moment and think about your company and the companies of your clients. Try to envision the situation. If only half the water you use now were accessible, how would it change your business world? Your community? What if the spring dried up and you had to rely on a fire truck to serve basic needs? What if you had to wear the clothes you have on for several days while "they" decide if you can have enough water to wash them? Don't worry, history says you'll survive. In the Middle Ages, even royalty wore the same clothes for a year.

TOO MUCH OF A NOT-SO-GOOD THING

Water management would not be complete without a few comments about controlling water when we have too much of it. Water poverty can also come from storms and flooding.

The massive Mississippi flooding of 1927 was the eye-opener to trying to manage excessive amounts of water. The flood covered more than one million acres with 30 feet of water. Five hundred people died, and more than 600,000 were displaced. Personal loss was staggering, and the economic impact was huge. It definitely changed the way the US government managed the Mississippi and other rivers. A comprehensive levee system was put in place, which now protects the area from the worst effects. Flooding today still does much harm, but it is managed more effectively.

The aftermath was almost overwhelming. One other statistic points to the huge management problems related to such a storm. It is estimated that the debris and devastation left behind

118 million cubic yards of "stuff" to clean up.

Earthquakes often cause tsunamis that can wipe out total areas. The more recent ones in Thailand and Japan reveal the immediate damage as well as the long-lasting effects.

Ironically, when citing these situations where there is too much water, it is well to remember that there can be a simultaneous problem of water scarcity. Flooding contaminates. Clean drinking water becomes highly prized.

GETTING ALONG WITH LESS

As you picture your world with limited water, consider that something as innocuous as those nice little sprays in the produce section of the grocery store designed to keep the lettuce crisp will be a luxury we can no longer afford. Then again, there's apt to be very little lettuce to keep crisp. At best, you could expect only a few shriveled-up heads to choose from. It takes a lot of water to grow food. Farmers will need to learn to grow more drought-resistant crops. When they do, we'll still need water to process and ship the produce. The message here is that there will be a far greater impact on your business life and your personal life than how much water comes out of the faucet.

Of the 6.9 billion people on Earth, 1.1 billion do not have access to safe drinking water. Another 1.8 billion do not have access to water in their homes. Add it up and 2.9 billion, or

40 percent, of the world's population…

does not have access to the water we take for granted. While they struggle to get five liters a day, we flush more than 68 liters down each toilet every day. Each toilet. Every day.

As shown in Figure 11-9, we use the most water per capita in the world—more than 656,000 gallons *per person* annually. Visualizing that, it is close to an Olympic-sized swimming pool PER PERSON.

So far, the majority of the water problems have been "over there," but the era of water scarcity is lapping at our door. Denial will no longer work. The water wealthy of one time, like Atlanta, are about to know the meaning of water poverty. We have rested on our laurels too long. Water complacency has been a way of life that we can no longer afford.

COMMAND AND CONTROL

The phrase, "command and control," was once a common way to describe governance in the former Eastern Block communist countries. I had the privilege of working beside some great people when the yoke of oppression was lifted. The relief was palpable.

Should a commodity as essential as water become limited, government controls approaching those similar to what the Eastern Europeans suffered is quite likely. Ask my friends in Bulgaria, Romania and Hungary, and they will tell you it should be avoided at all cost.

If we don't solve our problems locally, history tells us the federal government will step in. The past is prologue unless we learn from that past.

Once more the general public seems oblivious to the crisis on the horizon, only this time the water crisis will make the energy crisis of the 1970s look like child's play. In the 70s we reassured ourselves that "Yankee ingenuity" would be there when we needed it. It took nearly 40 years, but we seem to have been right. Between alternative energy and fracking, we have once again temporarily brought a crisis into manageable proportions. But where is the "solar" alternative to water?

There was a general feeling of "we're all in this together," during the energy crisis. Water scarcity will create a different mood. Problems will be more localized. We are more apt to have competition than cooperation. The acrimony will be greater and the lawyers will be fully employed.

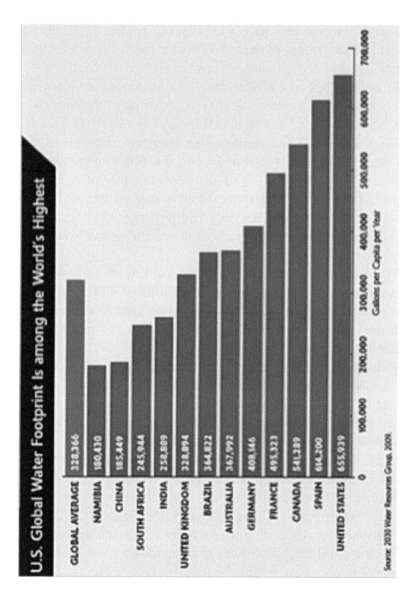

Figure 11-9. Relative Water Consumption per Capita

Our water crisis has the potential to turn neighbors against each other, communities against communities, states against each other, and even countries. A 21st century version of the range wars in those old western movies could be in our future. The economic implications are huge. Having NO water creates a price tag no one can pay, but basic health issues override even that.

In an attempt to call attention to this serious concern and its economic ramifications, I authored an article, "No Water is as Costly as No Water," for the AEE magazine, *Strategic Planning for Energy and the Environment*. The attention it received was gratifying, but it hardly put a dent in the public conscience.

It is not an exaggeration to say water is a matter of national security. Consider the impact if water is contaminated; it could leave a city or a nation in peril. The terrorist implications are horrendous. Just how well protected are the water sources in Washington, D.C.? Or New York?

What if you can't trust the water coming out of *your* tap? We cannot imagine the devastation to human health and economic prosperity if clean, safe water is not available to us. Such conditions can leave a city, a region, a nation, even the world uninhabitable.

Remember the lyrics, "Raindrops keep falling on my head?" Great song, but you might not own those raindrops. Remember those cisterns of yesteryear? In a world of water scarcity, it feels like a romantic off-the-grid notion worth trying again. Well, they are likely a thing of the *permanent past*. A man in Oregon has already faced the inside of a courtroom by trying to keep rainwater that fell on his property. More to come.

Probably the greatest harbinger of the water control fights on the horizon is the Great Lakes Water Resource Compact. The Compact, which came into force in 2008, was agreed to by all the states and two provinces that surround the Great Lakes. The agreement treats groundwater, surface water and Great Lakes tributaries as a single ecosystem. Anyone outside the Great Lakes basin, who wants to use the water, must meet strict criteria. Any diversions must be reviewed first by the states/

provinces and the Compact Council. Because the courts have ruled that states don't own the rivers that run through them, it can be relatively certain that the Compact will face considerable litigation over time.

A study of the Compact reveals some potential internal problems as well. Figure 11-10 has one striking feature—the percentage of water consumed by Ontario. When the chips are down, there could be some serious discussions with our Canadian friends over how the pie is "divvied up." If Pennsylvania suddenly has water scarcity, its occupants will be asking why their sliver is so small and Ontario has so much. It does not take a crystal ball to see this has the makings of an international dispute.

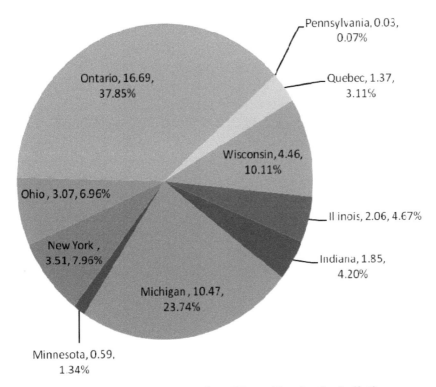

Figure 11-10. 2011 Great Lakes Water Use by Jurisdiction

CE CERO DORMIVO

When someone is suddenly taken dead through an abundance of bullet holes in Italy, and witnesses are questioned as to what happened, the traditional answer is, "*Ce cero dormivo.*" Roughly translated, it means, "If I was there—and I'm not saying I was—I must have been asleep."

When disaster looms large, do we do the Mafia thing and pretend to be asleep? Or, are we so oblivious that we really are asleep?

Those long gas lines of the 1970s, and our total ignorance of what energy conservation was all about, should have taught us a lesson. But we are lining up to have another crisis delivered to our door, and we are once again asleep at the switch. *Dormivo!* Time to wake up.

WATER IN OUR FUTURE

Singapore offers a global model of efficiency, as it operates a nearly closed-loop water management system, which supplies nearly 5 million people with about 60 percent of their water needs. Currently the island nation is dependent on Malaysia for its water, but its goal is to be self-sustaining within 50 years.

If we will only recognize the problem as Singapore has and plan accordingly, the future need not be so bleak. During the last 50 years through strategic planning as well as careful investment in research and technology, the country's water agency, PUB, has built a water system to meet Singapore's needs. The strategy is a four-pronged effort known as the "Four National Taps." As seen in Figure 11-11, the system includes (1) imported water, (2) local catchment water, (3) the purification of reclaimed water, which has been named NEWater, and (4) desalinated water.

The most important part of the Singapore story is that they recognized the problem a long time ago and are dealing with it.

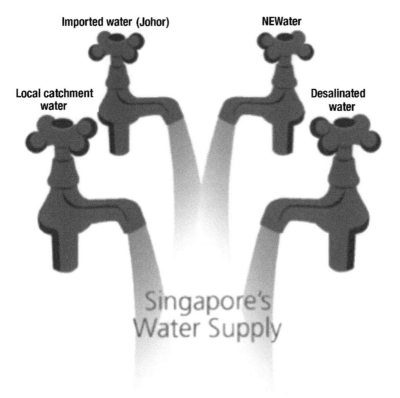

Figure 11-11. How Singapore Manages its Water Supply

In an era of increased concern about sustainability, there have evolved five environmental metrics. Those who are concerned about sustainability, automatically worry about energy efficiency and conservation, which includes concerns about carbon emissions as well as hazardous waste and waste sent to landfills. What is frighteningly sparse in these discussions is any reference to *water use.* Check the program for the much applauded 2015 Paris Conference.

Fortunately, environmental stewardship is becoming the watchword of many companies. The *Boeing 2011 Environmental Report* is a good example of efforts by companies to set goals and report on achievements toward those goals. Boeing reports water intake was reduced by 4.8 percent from 2009 to 2010.

If you are looking for the job of the future for yourself, your child, or your company, consider water management. The demands for expertise in this area will be staggering. As noted in Chapter 1, necessity is not only the mother of invention, but the mother of expertise. The learning curve is a steep one, but the opportunities are huge.

Land near water has been described as the Saudi Arabia of the future. If you want to invest in land with a future, consider those reservoirs we call tributaries, bays and seas. For those of us fortunate enough to live on the West Coast, we have a huge reservoir called the Pacific Ocean at our doorstep. Just wring out a little salt, and the supply is virtually limitless.

A tantalizing thought: Look at the huge resource opportunity in marine energy. This emerging field offers a great opportunity to draw energy from the same water we are treating. Opportunities in this area include wave power generation, tidal (current) stream technologies, salinity gradient power generation, and thermal gradient.

Each of these areas has a raft of energy capture ideas. For example, in wave power generation, work goes forward in wave capture devices, shoreline devices, oscillating water columns, offshore wave energy converters, floats, wave pumps, etc. Exploring salinity gradient includes cosmic power, hydrostatic generation, vapor compression and reverse electrodialysis.

As we pursue these opportunities, we must be mindful, however, that roughly 90 percent of life on Earth is in the water. New species, especially those found at great depths, are being discovered every year. It is critical that the pursuit of marine energy be done in harmony with marine life. Small mistakes could cause serious damage as well as furthering resistance to additional important research.

The threats associated with a water crisis are almost overwhelming, but they can be managed if we wake up and plan accordingly. We don't need another "Oil Embargo" to wake us. Water availability will determine where we live and how we live. We can pay attention now, build our future accordingly, and arrive at a new life style rather painlessly. Or, we can continue to look the other way and suffer another big resource shock in the not-too-distant future.

While water shortages are popularly ascribed to climate change, a *New York Times* article, "Southeast Drought Study Ties Water Shortage to Population, Not Global Warming," tells a different story. The article summarized the work of a Columbia University researcher on the subject of the droughts in the American Southeast between 2005 and 2007. According to findings published in the *Journal of Climate*, water shortages have resulted from population size more than rainfall. Census figures show that Georgia's population rose from 6.48 to 9.54 million between 1990 and 2007. After studying data from weather instruments, computer models, and tree ring measurements, they found that the droughts in the Southeast were not unprecedented and resulted from normal climate patterns and random weather events.

Chapter 12

On the Waterfront

Water poverty is not one-dimensional. It comes in many shades of desperation. Lost jobs, hunger, communities which no longer function, growing health concerns, violence and even death can all result from water scarcity. The pain and suffering can range from worries to stark tragedy. The emotions and confrontations that come with water poverty can range from a sense of disbelief to acrimony or even war.

According to Dow's Water & Process Solutions division, by 2030 with average economic growth and current usage patterns, our water requirements will grow from 4,500 billion cubic meters to 6,999 cubic meters.[12-1] That will be a 50 percent increase in just two decades. Analysts predict that available water supplies will satisfy only 60 percent of demand as population, urbanization and higher living standards continue to rise. By the time today's 3-year-olds graduate from high school, our water supplies will satisfy *only 60 percent* of our needs.

Shades of desperation will certainly vary by locality. They will also vary significantly by how soon we recognize the problem, how we go about making people aware of it, and how effectively we plan and implement that plan. There are many unknowns in predicting water scarcity, but it is absolutely certain that having no plan is an invitation to disaster.

PLANNING OUR FUTURE

Coming to grips with the myriad issues we might face is certainly no walk on the beach. Unlike energy, water problems are local; so the value of certain solutions to a specific problem

will vary. The issues discussed below, therefore, have no ranking. They have deliberately not been numbered to underscore
this fact and are drawn from our experiences with the energy
shortage for your consideration.

The best solution for a certain locality may not even be
mentioned in this discussion. In fact, we may yet have to dis-
cover it.

What we regarded as brilliant in 1978 to address our energy
needs may turn out to not have any relevance at all to our water
worries. We are, however, dealing in a commodity, and we are
dealing with people; so some striking similarities are likely. The
intent is to look at pieces from our energy experiences and then
determine how they might come together to develop plans that
fit *your* water needs.

WATER EXPERTS

Chapter 1 emphasized that we can create our own experts.
They need not have a specialized background in water. They
need not have a science background. There are some good books
to help establish a reference base and an increasing number of
seminars on the topic. Unfortunately, the seminars seem to be
following the energy management path of focusing on technical
matters. In most cases, that will be the easy part. Some strong
leadership expertise will be desperately needed.

As the energy crisis emerged, nearly all our focus was on
the technical aspects. Energy managers were seldom treated like
managers. Even now, as vital as energy is to most companies, it is
a struggle for energy managers to get a seat at the management
table. It would be a serious mistake to follow our energy crisis
pattern and go forward as if water scarcity is only a technical
issue.

To resolve our water concerns will require a change in behavior in most of our populations. This will empower people
who take the time to become informed and *who can lead.*

Water expertise will be in increasing demand. Getting people to take the lead NOW in a company, local government, and/or state government will very likely pay big dividends later. Those who cut their teeth in the energy field and are hungering for a new challenge are apt to find some lucrative opportunities as our water problems become more prevalent.

DENIAL

It's surely a good thing that there are a lot of sandy beaches on the waterfront because it's going to take a lot of room to accommodate everyone who has his or her head stuck in the sand. Way too many think a "good rain" will solve any drought problem. Turning around the entrenched denial will not be easy. In fact, getting people to even acknowledge that we have a problem will be a huge challenge.

As you followed the path of our energy efficiency evolution, I hope you got a sense of the mad scramble we experienced. Even though our relations in the Middle East were becoming fractious, we were not prepared for them to turn off the spigot. Remember those chickens with their heads cut off? Well, we seem to be laying the groundwork for a lot more chickens.

Early on the docket is a need to get our water problems front and center. The public must be educated as to the imminent water crisis we face, and our need to understand the ripple effect it is apt to have around the nation and the world. For example, a drying aquifer in the Midwest will effect employment, food supplies, and the cost of food. With no water, an entire area now densely populated could become desolate. The ramifications could be huge. Putting it off will only cause more pain, more cost, and more chaos.

The epitome of denial is shown in the behavior of some of the farmers that rely on the Ogallala aquifer. Instead of looking for ways to preserve the dwindling water supply for future generations, they are reportedly increasing irrigation dramatically. These actions seem to be governed by a "get all we can while the getting is good" mentality. It's hard to tell if they are inspired by greed or bogged down in stupidity. In fairness to the farmers/ranchers, however, it is hard to stare at a dry field and not give it water.

In either case, it sounds like the government needs to step in with a carrot/stick approach while there is still time. If we don't wake up and recognize what we are doing to the Midwest, we could have the likes of another Dust Bowl from the 1930s. Figure 12-1 is a stark reminder of what those horrible dust storms did, and we don't want to go there again.

Figure 12-1. 1930s Dust Bowl

To overcome the apathy, most localities will need an awareness campaign with teeth in it. We need to go beyond making people aware; we need to provoke action. A sense of immediacy needs to be laced into the message. We are straddling a fine line between motivating people to act now, and creating the elements of a panic; so some caution is advised.

While constructing our message, we need to be very aware that water is essential to every living person, and they are going to be emotional about the situation. Shortages can create violence. Our energy shortages in retrospect were relatively calm compared to what people may do when fighting for survival, or even *perceived* survival.

The challenge will be to create a message that alerts people to the imminent problem while offering reassurance at the same time. Leaders will need to assure people that something is being done about the problem. Of course, it would be a very good idea if they know what that something is.

We can learn from the effective actions in some cities/counties, which have addressed local water problems, such as Las Vegas, or states like California and Oregon. We can also learn from their mistakes, such has the historical denial in Atlanta. We do need to be cautious, however, because remedies in one locality may not work in another.

Our propensity to look the other way is strongly supported by our political system. Our elected officials have no incentive to bring harsh realities and adverse conditions to the attention of their voters. Especially if the solutions mean some measure of deprivation in the short term—even if some clear benefits are evident in the long term. There are a host of short-term negatives and long-term solutions. The inducement to act in a way that causes short-term pain is simply not part of the political process. Conversely, the temptation to ignore the situation is often rewarded.

GOOD BUSINESS

Successful businesses always keep an eye on needed resources. Water scarcity will not be impervious to supply/demand pricing. We have been lulled into complacency. It is just good business to do the local research and anticipate such an event. Protecting your resources and then using them as cost-effectively as possible is not new to good business practices. Water is no different; it's just that we have not paid sufficient attention until now, and our efforts may need to be more strident.

In fact, a business may be in a position to alert the community. Businesses that take a wait-and-see posture may pay a hefty price tag down the road.

Those who are already making a concerted effort to cut reliance on water in their processes, like Boeing, will be a step ahead. When the government mandates start flying, however, it's critical that we don't have across the board cuts that will, in essence, punish those who were out in front of the effort.

Every business sustainability plan should have
a strong water management component.

While corporate investment in water and wastewater was up by 53 percent during the last half of 2013 and first half of 2014 for a total of $140 million into 33 ventures, it did not reach the dollar count for the first half of 2012, which was at $317 million. Despite the relatively low amount of venture financing in the water sector, companies ranging from 3M to Opower have entered into partnerships with start-ups such as Water Planet Engineering and Water Smart Software, respectively.

COMMUNICATIONS

Implicit in many of the concerns listed here is the critical thread of communications, but it warrants mention on its own. It is almost impossible to overstate the importance of effective communications. It is also valuable to use pictures. The old adage that a picture is worth a thousand words still holds true even in our high tech world.

Every water plan (or sustainability plan with a water component) needs a communications strategy. A key ingredient in that plan is someone designated to be responsible for commu-

Figure 12-2. Water Scarcity Landscape

nications. If everyone is responsible; no one is responsible, and it does not get done.

All too many people think they are effective communicators; so a "strategy" seems unnecessary to them. When a communication effort relies on "I see Joe almost every day, so I'll let him know," the word doesn't get out like it should.

Since emotions are apt to run high during water scarcity, some special sensitivity as to exactly what is written or said is important. We also need to be aware of words that trigger emotions.

Being aware of some proven strategies and cautions may prove critical.

1. **Know your audience**. There will be some surprising variations in receptivity, interest and even acceptance. In an organization, the basic message may be the same but it will need to be tailored as to whether it is going up, down, or sideways. The same wording or even the same approach is seldom equally effective in the board room and the boiler room.

2. **Timing**. People are ready for a message at different times. If your company is in the middle of union negotiations, a message about a coming water shortage can be easily distorted; or, it may be totally lost in the shuffle.

3. **Content.** How immediate is the problem? Content is often time sensitive. The city officials in Flint seriously missed informing the public of what they knew about contaminants in the water.

There is always a question of how much should be shared. The best guidance is to tell them what they really need to know and stop. Even in a broader context, wording or the use of certain graphics might change how the message is perceived and received.

Be aware of the impact your pictures, or others, might have.

As much as you might like golf and understand perfectly why the greens need to be maintained, this picture of a golf course will be a real turnoff to some people.

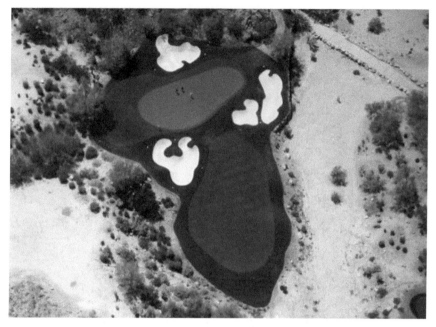

Figure 12-3. This is a *nice shot!* It covers fore *and* aft

Be aware of other concerns that may be impacting your audience. Time your presentation for maximum receptivity. Or, if you want to downplay the situation, you can take a page from the federal government and release your statement late on Friday—a holiday weekend is even better.

4. **The message.** The number-one rule is to be comfortable with your message and stick to it.

   Crafting a message is not a simple task, but simple wording is often most effective. Watch the inclination to do a "data dump." As you learn more about your situation (as you really get into it), resist telling the world everything you know. Tell them exactly what they need to hear and no more. What people most want to hear is a few words as to what the situation is, why now, how it came about, how it will affect them, and what you are doing about it.

The debate goes on as to how effective signage is. It's hard to measure how much a sign like the one in Figure 12-4 motivates people to use less water.

Figure 12-4. Serious Drought Signage

In emotionally laden times, it is well to remember that there are those who are frightened, trying to fight back, or see an opportunity for political gain. Given the slimmest chance, they will twist your words. Again, avoid words that are apt to trigger strong emotional response. Use the "triggers" only when you are trying to get them to act or change behavior, and then do so cautiously. Go over what you write or intend to say with a "fine-tooth comb." Examine the phrases and make sure you are not providing fodder for the protesters. Then, get another pair of eyes on it. Avoid descending into a war of words.

It is absolutely critical that the statements be accurate. Be careful in naming names when citing the history of the problem. Use only verified numbers. If something is an approximation, say so. Communication strategy in times of disaster is an art in itself. There are books, lectures, and public relations firms out there to help. In the interim, a few admonitions seem in order: (1) be brief and stick to the facts; (2) control the background—on a TV interview, they will try to put you in front of the disaster, (3) never predict, and (4) never, ever respond to hypothetical questions.

Accuracy relies on careful assessment of the prevailing conditions. It is often the better part of valor to say, "I don't know," or "We are currently looking into it and I'll get back to you." An audit of the situation, including the exact water source and usage is critical. In a closed-door session, it's important to consider the ramifications of the data before you; but, once in a public setting, do not predict, do not speculate.

Be aware of the sources of reliable information. The map and accompanying discussion in Figure 12-5 represent just one piece of information available from NOAA Climate Prediction Center.

Watch for myths and misleading information. For example, much is said about hydroponics (growing crops in water) consuming a lot of water. In fact, hydroponics saves between 70-90 percent more water than soil, as water is recirculated and reused. The crops may yield up to three times that of traditional gardening.

When it comes to numbers, accepted standards of measurement are essential. We all need to be using the same yardstick.

Above all, it is well to remember that in communications, silence is fertile territory for growing fears.

MEASUREMENT

As we gradually emerged from the energy crisis, we found we needed some universal agreement on how energy savings should be measured. It took several tries to get it right, but even-

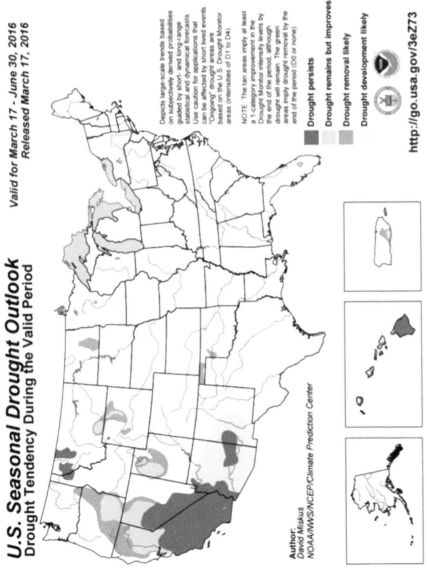

Figure 12-5

tually International Performance Measurement and Verification Protocol (IPMVP)[12-2] emerged and was internationally accepted. Having served as the first president of the IPMVP, I had the privilege of working with an incredible group of people who put in a horrendous amount of work to give us a quality protocol. It is an on-going effort, and refinements are being made even today.

THE IPMVP is an excellent body of work and is a good model for developing a comprehensive water assessment. The work has addressed water to a degree, but it needs to be more fully developed. Eventually, a stand-alone document may be needed. Water measurement will need to take into consideration elements in the community or region beyond those considered in a survey of potential energy savings.

Working with the Association of Energy Engineers, the IPMVP leadership also created a certification program to establish which professionals were qualified to do the measurement and verification. This gave credibility to the program and the process. Further, it offered energy service companies the needed assurance to support transactions. It also gave the owners the assurance they needed in getting outside assistance. On a third front, it gave financiers a way to gauge risk and a level of reliability they were looking for.

The more heated the discussion and the resistance, the more critical it becomes to have credible data. A water credentialing process will prove to be very valuable.

When it comes to measurement, we have another déjà vu all over again. Just as early versions of the IPMVP lacked a baseline, we are having trouble establishing a baseline for measuring progress in the water world. *The Economist's* 2016 edition of the "Pocket World in Figures," under Country Profiles addressing Health and Education, presents some precise percentage figures of "improved water source access." The pocket guide tells us that Cameroon is now at 74.1 percent while Kenya is 61.7 percent, but compared to what? What does "improved" mean? What were the conditions before? Does "improved" mean that they still walk over a mile to get water, but they now use a bucket instead of a teaspoon to collect it?

If a baseline exists, a report should share with the reader what that baseline is and the criteria used to establish it.

If we are to effectively communicate the extent of water problems around the world, we need to establish some baselines as well as universal protocols and procedures. If we are to make the case, we need to be working from the same playbook.

We need a common yardstick
that everyone understands.

Water meters have been around for ages; so you may think the measurement process is all taken care of. Have you ever wondered just how reliable those meters are? Talk to people in the business, and they will tell you that those meters gradually deteriorate, and as they do, they lose accuracy. Interestingly, the growing errors in water meter measurement usually favor the supplier. In an era of cheap and readily available water, we haven't cared too much. Those carefree days, however, may soon be in your rear view mirror. As prices climb, a new meter just might be a good investment.

Not surprisingly, water analysis people have come up with one other measure, the water footprint. It is envisioned that the water footprint will be used to make comparisons similar to the use of the carbon footprint. At best, it is most likely to be a generalization of how much water a company, community or country uses. It will call attention to how much is being consumed and prompt the more judicious use of the commodity as it becomes scarce. If used too casually, however, the water footprint can be dangerous, as when the environmental activists threatened to boycott the roses being shipped to Europe from Kenya because the rose growers' water footprint was "too big." Further investigation revealed that it was the most efficient use of water to keep the industry in Kenya going and the farmers employed.

Water will become an even greater political tool, just as

oil has. The water footprint will become a bone of contention. It is apt to be used by those with a smaller footprint to change practices in other countries where water is more plentiful.

One other type of water that will eventually need some quantification is virtual water. When any product, such as food which requires considerable water to grow, is shipped to a drier area, the water used in growing the food is referred to as virtual water. It is, in effect, imbedded in the food. Shipments of food and other products from water-plentiful to water-scarce areas have been going on for decades, but as water scarcity grows, the process will come under closer scrutiny, and some means to quantify this virtual water should be expected. This, as with most resources, is most apt to be reflected in the price of goods. The difference is that we will be more conscious of it.

Great care needs to be taken when using the water foot-print to compare various entities because there are many other variables at play. For example, South Africa, which is relatively short on water, currently exports virtual water because it has the financing, technology, infrastructure and institutions that its neighboring countries do not have.

However we measure it, the process has far to go. Morgan Gillespie of the Carbon Discussion Programme recently reported at a meeting in Johannesburg that 30 different water assessment tools are being considered by the program's members because there is no agreement as to what should be measured.[12-4]

Because water gets measured in many ways; i.e., liters, gallons, pounds, there is a list of equivalencies in Appendix A.

MARKETING

Closely aligned, but separate from communications, is the need to consider the marketing of the concept. Telling the story may be prompted by the need to change use patterns, the need to reuse water, or the installation of a desalination process.

Despite all its benefits, we found energy efficiency a hard

sell until we really came to understand the market. Selling water efficiency and effective water management practices is apt to meet with initial resistance as well. In the US, management is accustomed to having cheap, available water. While it is essential to life and business operations, management takes it for granted. Typically, operations and maintenance people only think about water in terms of plumbing problems or process procedures. Capturing their attention and convincing them of the need for advanced planning will demand some good marketing skills.

Capturing wastewater and reusing it will be a whole new concept to many. Overcoming apprehensions with clear explanations and solid data will be key.

If we draw on our experiences in energy, we are most apt to get management buy-in when we talk money. Under the technology discussion, two new technologies that pay for themselves in a short time are presented. The economics are attractive and might be a good place to start.

As we did with energy, we are going to need the operations and maintenance (O&M) people's involvement. Project success will depend on the O&M people seeing why they should be involved and how the changes will benefit them, their organization and the community.

With both management and O&M, we must remember to fish from the fishes' point of view. It's just good marketing.

The marketing strategies may need to embrace a public relations campaign to overcome uneasiness or fears related to shortages or water reuse. This will be particularly demanding if there is an outbreak of a communicable disease after water reuse is implemented. Once again, there will be troubling tangents, similar to our indoor air quality concerns. Expect fear, confusion, misinformation, charlatans trying to make money out of the deal, and protestors with concerns that may not even be related to the matter. Take a page from the energy efficiency/ IAQ-related problems, insist on hard data, and avoid conjecture. Insist that consultants speak in simple terms. Do not be impressed by new jargon or medical terms. Cut through the jargon and insist on solutions. If aerosols from used water are suspect, get tests immediately. Be sure the applications are appropriate; i.e., watering outside with a drip rather than a sprinkler.

WATER AUDIT

The most effective water audit is not going to match the footprint of an energy audit. If we try, it will create a lot of false motion. More importantly, it's apt to overlook some key factors.

An energy service company, which just finished a water project in another country and issued a report recently, referred to its assessment as an energy audit. We need to shift our thinking. Yes, there will be some similarities when it comes to technical considerations, source options and availability concerns. It is absolutely certain that, like an energy audit, the audit alone will not reduce consumption. If we don't act on our findings, we will not achieve the needed savings. Water audits can gather dust just like some energy audits have.

A water audit must put even greater focus on the people factor. It will need to be a step up from the investment grade audit. Since an audit is the foundation for an effective plan, the most telling aspect of the audit is apt to be the leadership qualities of management and the extent to which the organization will lend itself to changes in behavior. Whether we are talking about a government entity, a non-profit or a business, an assessment of the "corporate culture" will be a key part of the audit.

The technical aspect of the audit must assess the amount of water coming onto the property and the amount actually consumed. Underground leakage is a common problem and must be detected. Resource options will need to be assessed, improved efficiency opportunities identified, and the economical benefits calculated. There will be an overlap with the energy audit because water needs energy to be used. Methods for heating and distributing hot water will need to be considered.

The new challenge in many audits will be the assessment,

acquisition and installation of a reuse system, including the process itself and any needed double plumbing.

Arriving at a consensus of what a water audit should be is crucial. Both content and process must be established—and soon.

TECHNOLOGY

In emphasizing the people factor, which is too often given a short shrift in energy audits, the impression may be given that the technology is not so important. It is, of course, a key factor. Staying on top of new technologies as they emerge is essential.

One of the good news stories of late is the work being done by the Dow Chemical Company. Because desalination and water reuse are such vital parts of extending our water resources, the technologies associated with salt removal and turning low-quality wastewater and raw water sources into high-quality water will become an essential part of sustainable economic growth. Dow's Water & Process Solutions division has invented a new breakthrough in polymer chemistry, resulting in a new family of reverse osmosis products. Compared to the existing membrane, the new product filters out 40 percent more salt from water while consuming 30 percent less energy. [12-3]

A German metalworking company reports that the use of a vacuum distillation system produced by H_2O GmbH has reduced the amount of wastewater produced by 98 percent and paid for itself in just one year. The system evaporates 100 percent wastewater to produce 98 percent pure water, leaving only 2 percent residue. The producers of Vacudest claim that any company generating more than 180 cubic meters of wastewater each year can save money by installing this distillation system.

Once in awhile an innovation comes along that brings a big smile to the face of an energy engineer. Lucid Energy has come up with one. If you have water flow in a city that goes

Figure 12-6. Lucid Energy
(Photo: Keely Chalmers, KGW News)

downhill, the turbine produced by Lucid Energy will capture the energy and put it to use.

In Portland, Oregon, the city now has water pipeline producing enough electricity to power more than 100 homes. Drinking water flowing through a water pipe below Southeast 147th Avenue and Powell Boulevard spins four turbines.

That spinning creates energy for PGE customers. Even better, this technology does not pose any environmental damage. No fish are hurt. No dams are needed. It has no effect on the surrounding area.

This pilot project cost USD 1.5 million, and is being funded by the manufacturer, which is sharing the revenue from the electricity sold to PGE.

"Energy is the single large cost, so if we can recapture energy that's already embedded in the flow, we can reduce the cost for everyone who needs it," said Gregg Semler, president of Lucid Energy.

The advantage of the Lucid pipe system is that it produces electricity all the time, around the clock, without any environmental impact.

As we have learned from our energy experiences, innovation can help solve the world's problems in a more sustainable way. Cost-effective new technologies need to be part of audit considerations and water management planning.

FIT TO PURPOSE

Water is not equal in value or purpose. At the moment, we often use high value potable water where a lesser quality would serve the purpose. Since high value drinking water is cheap and readily available in most of the US, we squander it. As we move forward in our efforts to manage water more effectively, we will need to examine more closely exactly how we use water. One key aspect of an audit will be to survey how closely use fits the purpose. Gradually, we will become more conscious of the categories of water and how to use each to the best advantage. The audit will also need to address the procedures for managing each category most effectively. As we become more sophisticated at the residential level, for example, wastewater will not go down the drain into a sewer, but will flow into a holding tank. The tank may even have some simple filtration. That reuse water will then serve several purposes, such as washing the car, watering the garden, or flushing the toilet. Greywater will become part of life. One system is shown in Figure 12-7.

Manufacturers will examine reuse potential and find more effective ways to recycle water in their processes. Because 49 percent of domestic water consumption is used by utilities, finding ways for utilities to reuse water could also ease the scarcity in certain localities.

As we more carefully define water by its purpose and quality, its price may vary. As the price of potable water goes up, we'll have an added incentive to use it over and over again.

Figure 12-7. A Delayed Distribution Greywater System

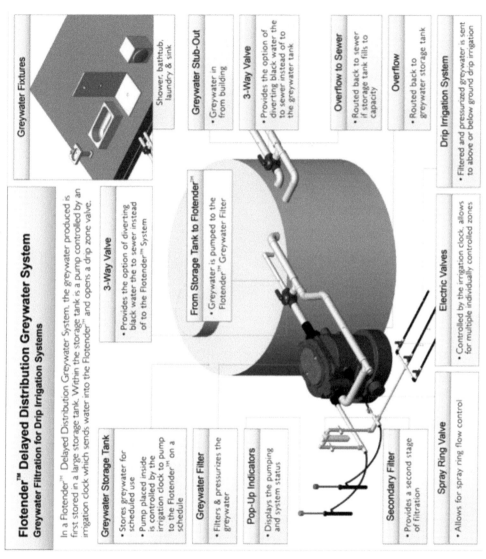

Flotender™ Delayed Distribution Greywater System
Greywater Filtration for Drip Irrigation Systems

In a Flotender™ Delayed Distribution Greywater System, the greywater produced is first stored in a large storage tank. Within the storage tank is a pump controlled by an irrigation clock which sends water into the Flotender™ and opens a drip zone valve.

Greywater Fixtures
- Shower, bathtub, laundry & sink

3-Way Valve
- Provides the option of diverting black water the to sewer instead of to the Flotender™ System

Greywater Stub-Out
- Greywater in from building

3-Way Valve
- Provides the option of diverting black water the to sewer instead of to the greywater tank

From Storage Tank to Flotender™
- Greywater is pumped to the Flotender™ Greywater Filter

Overflow to Sewer
- Routed back to sewer if storage tank fills to capacity

Greywater Storage Tank
- Stores greywater for scheduled use
- Pump placed inside is controlled by the irrigation clock to pump to the Flotender™ on a schedule

Overflow
- Routed back to greywater storage tank

Greywater Filter
- Filters & pressurizes the greywater

Drip Irrigation System
- Filtered and pressurized greywater is sent to above or below ground drip irrigation

Pop-Up Indicators
- Displays the pumping and system status

Electric Valves
- Controlled by the irrigation clock, allows for multiple individually controlled zones

Secondary Filter
- Provides a second stage of filtration

Spray Ring Valve
- Allows for spray ring flow control

Figure 12-8. A World of Reservoirs

WATER "ALTERNATIVES"

There is no solar solution to our water problems. We can use efficiency and reduce water waste, and we can repair our infrastructure. However, the available alternatives to ease the energy shortage, such as solar and wind, don't exist. Our only way to augment what we have is to make unusable water useable. There are two source possibilities: (1) we can reuse our wastewater; and/or (2) we can desalinate water from the oceans. Desalination is relatively expensive but doable. There are desalination plants all over the world, and we are apt to see a proliferation of more. Our oceans offer incredible reservoirs. As with energy, we may have some NIMBY problems when we try to lay the pipes to get the desalinated water from the plant to the consumer.

Reuse will be more economical, but its application limited. Wastewater storage and double plumbing are apt to become the new normal in many locations.

SUSTAINABILITY

Water management should be the cornerstone of any sustainability program. Without water, we have nothing to sustain.

Water management can be a stand-alone effort if an organization does not have a cohesive sustainability plan. Water scarcity might be the impetus to make a broader sustainability effort.

A sustainability plan that does not address water concerns and efficiencies is not complete. In fact, it misses the essence of what sustainability is all about.

PRICING

As indicated in the "Fit to Purpose" comments above, the future price of water is apt to vary with the quality as well as availability. As it becomes scarce, the price is apt to go up with its perceived value. The paradox here will be that water straight from the babbling brook will probably be cheaper to supply than the water that must be treated. Some means of passing on the costs of cleaning water for reuse and desalination must be agreed upon.

A strong component of managing a commodity is its pricing. There is ample evidence to suggest it is the most effective means of managing a commodity.

Since water is an essential commodity in our lives, the pricing must not become prohibitive for the poor. Some provisions for federal or state support will most likely come into play.

The pricing of water will be an important aspect of the planning process. The consequences of certain price structures will need to be carefully considered. Hopefully, this will be done locally. Price regulations that do not consider local conditions could do more harm than good. On the other hand, if subsidies become part of the picture, some control might be necessary.

CREATING THE PLAN

A whole manual could be written on water master plan development. Here we touch on just a few considerations and then list essential ingredients of a plan.

Water plans do not function in isolation. Water management is part of a bigger picture. How the audit recommendation will interact with the overall operation and human behavior must be considered. Broader cost implications need to be factored in. For example, what if the water audit recommendations negatively impact the process or require other changes that are prohibitively expensive? Cutbacks in water availability may be the deciding factor, but in most instances, trade-offs may well require management to do a balancing act.

A water plan must be based on solid information; so a quality audit is integral to the plan.

Every plan will be unique to the local situation and circumstances. Some common components with no priority are:

a) Explicit information on the source(s) of water by quantity and quality currently available; options for additional sources with potential amounts, availability, and associated costs.

b) Testing procedures, recommended testing intervals, and specific details regarding timely reports of results.

c) Measurement and verification procedures (of quantities, not testing) need to be established.

d) Codes/standards and compliance provisions.

e) An inventory of existing equipment and piping detailed with graphics; needed modifications and efficiency improvements; needed replacement and procurement procedures.

f) Opportunities to use the water more effectively and efficiently.

g) If a reuse system is in place, it should be evaluated, including storage, process, testing, distribution, and ability to meet current needs; modifications needed and cost.

h)   If a reuse system is not in place, the audit should examine the feasibility (both technically and economically) of installing one, and the findings should be part of the plan.

i)   Water security where appropriate (some concerns may be classified).

j)   Emergency procedures in the event of contamination.

k)   Associated operational implications; i.e., more or less energy required for new system.

l)   An assessment of current use by function and groups will be critical if water scarcity is a concern.

m)   Environmental considerations.

n)   An overall communications strategy regarding routine information and responsibility assigned; special attention to procedures during distressed conditions should be clearly set forth and specific responsibilities assigned, including sign-off regarding announcements and press releases.

o)   Provisions for the initial audit and periodic reassessment.

p)   A schedule to evaluate and update the plan as warranted.

q)   The position of water manager should be established—its position in the organization and the power of the office.

r)   A water policy.

During water scarcity, hard choices will need to be made. The policy should state the criteria and procedures that will be used to allocate water by category. Also a statement regarding how water quality will be categorized, allocated by group and its pricing should be clearly set forth.

Graphics in your plan and in your marketing, such as the one shown in Figure 12-9, will help people better understand what is contemplated and the safeguards to be used.

WATER AND POLITICS

This is a scary one. Having worked in 38 countries, I am more convinced than ever that we have the greatest country and absolutely the best political system. We are, however, not even close to perfect. In times of emergencies and disasters, our fabric is tested.

Nothing will test it more than an *unexpected* scarcity of an essential commodity, water. The term "unexpected" is used advisedly because many of us know the problem is already here in some localities and on the horizon in others.

You can, however, count on many politicians to treat it as "unexpected." Our elected officials, whether they are federal, state, county, city, or village, all want to be re-elected. If a politician makes a decision, some people (voters) aren't going to agree. So it is human nature to put off those decisions whenever possible. Even better, many have learned to look the other way and have convinced themselves no decision is needed; they have denial down to a fine art. A good example is the farmers in the Southwest increasing their irrigation using water from the depleting Ogallala aquifer. The farmers (voters) could be stopped by passing a law or issuing regulations to benefit future generations (not yet voters). Most likely the short-term resolution will be paying the farmers to use less water. Once more, the taxpayers are apt to foot the bill because it is unpopular to say, "No."

Under the glare of the problems in Flint and reports from *USA Today*, expect the EPA to come out with regulations on the safe levels of lead in drinking water. Given the propensity of government to be cautious, the new level is apt to be at the current "action level" of 15 parts per billion or below.

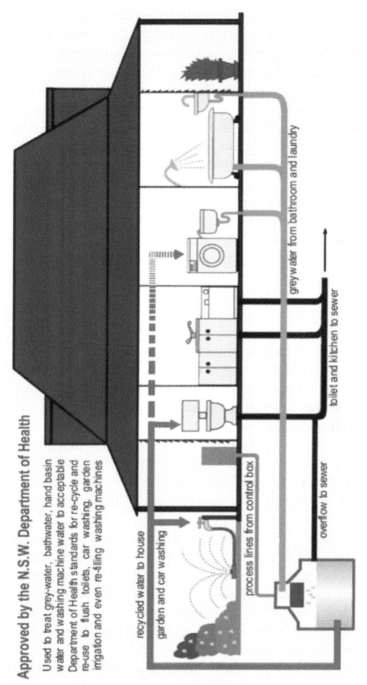

Approved by the N.S.W. Department of Health

Used to treat grey-water, bathwater, hand basin water and washing machine water to acceptable Department of Health standards for re-cycle and re-use to flush toilets, car washing, garden irrigation and even re-filling washing machines

recycled water to house

garden and car washing

process lines from control box

overflow to sewer

greywater from bathroom and laundry

toilet and kitchen to sewer

Figure 12-9. Don't Talk of Water, Show Me!

In all fairness, governments sometimes gets things right. Case in point: Decades and decades ago, the people in what is now New Mexico came up with a successful way to share the Rio Grande waters. They came up with "Paper Water." Yes, that is what it's called. At pre-ordained times, water is released to flood valley property. The owners know it is coming and how many acre feet they can get at a given flooding. They even have "ditch riders" who are out there making sure everything goes according to plan. It works and has for a long time. Like all such things, it is also in the courts. Since the ditches were built as early as 1844, the local people think that the federal government's control is not appropriate. At the root of the lit-igation regarding the Lower Rio Grande Basin is a process to determine the precise nature of all water users' claims to both underground water and river water. The contention has gone on for decades, and no end is in sight. In the meantime, at the local level, the owners get their "acre feet" and the ditch riders do their riding.

The greatest danger outside of New Mexico, when it comes to politics and water, is the likelihood that local politicians will put off what they should do and, at first sight of catastrophe, the state or federal government (or both) will march in with blanket mandates.

This will be compounded by the tendency that politicians have (once the catastrophe is on their doorstep) to "view with alarm" and rush to solutions. The fact that water poverty can be unique to a local situation and needs local resolution will totally escape them.

The "past is prologue." The pattern exists: The vociferous few trumpet a certain solution, and the uncertain politicians, who have not done their homework and are eager for a solution, will apply the idea in broad sweeps. Decisions will be made in far-off places, like D.C., by decision-makers, who are short on facts and have never set foot on the local stomping grounds. This might sound bitter, but it is reality.

Outreach has begun. The EPA is already playing with measuring the amount of water used in hotel-room showers. Any thoughts on where that might lead?

Political reality makes it imperative that local officials in any area with potential water scarcity act immediately to get plans in place. The plans may not be perfect, and they will need to be refined as the local people get more involved; but, officials must start thinking about it.

If the local folks have things under control, then there is less danger of widespread hysteria and eventual catastrophe. In such circumstances, most state and federal people will be relieved—no tough decisions to make. Of course, there will be those trying to build a bigger department that will love to wrap their arms around your problem and "help." Be assured, you do not need that kind of help.

Given our history, chances are very good there will be a fight among federal agencies over who can best take care of the little people. They will issue regulations trying to outdo each other. With the potential for high emotions and voter outrage, a chance to carve out a big piece of the turf will be irresistible. (Remember that federal pay grades *and salary* are largely dependent on how many people a government official has working for him or her.)

Let's fervently hope that we don't have an outbreak of flu or some communicable illness in a community which has just started using reuse water. It will make the indoor air quality fears associated with energy efficiency seem like small potatoes.

We are all in this together, and if the folks down the road don't get their act together, we'll all pay. *This is an excellent time for the associations, which serve cities and counties, to launch an awareness/education program.* For your own sake, you should insist that your association(s) do so.

> *If you have ever tried to build on something*
> *that even resembles a wetland, you have a hint*
> *as to what is in your future.*

It may help to recall that Albert Einstein said, "In the middle of the difficulty lies opportunity."

The choice for local leaders facing water scarcity is really quite simple. They can assess the situation, secure solid water-audit information, create a plan, and begin implementing it. Or, they can expect state, and more likely federal officials, to run in, take over, and never let go.

At this point, it seems only fair to share with you some good things government has done. The following are two brief case studies of a country, Israel, and a city, Las Vegas, and its surrounding territory. They not only prove that government can get it right, but offer some incredible ideas that might be useful to you in your own planning. These are only the briefest of summaries, but you are encouraged to study them in detail.

CASE STUDY: ISRAEL

In the short span of a decade, Israel went from water scarcity and fear of drought to an abundance of water. These efforts grew out of nearly 70 years of work by some fine engineers, scientists, and policymakers. Together they developed Israel's water-related expertise, technology and infrastructure.

While much of what Israel accomplished will not apply everywhere, there are some great lessons here for people trying to carve out a life in the desert. The temptation is to go to a checklist, but

the overriding factor of success was the change in culture.

It was first stressed that the water belonged to everyone. Then prominent advocates stepped forth, gained media attention, and created public awareness. The crucial turning point was creating a water-respecting culture. The message penetrated classrooms, supported by signage along roads and clear explanations of why water restrictions were needed.

Other than the obvious, such as water purification, desalination and rain capture, the country has encouraged drip irrigation for most of its agriculture, reformed agriculture to grow drought-resistant crops, priced water to encourage efficiency, financed water-saving R&D, and even developed seeds that thrive with salty water.

CASE STUDY: LAS VEGAS[12-5]

Water features dominate Las Vegas from the 1.6-million-gallon Shark Reef Aquarium at Mandalay Bay to the spectacular fountain at the Bellagio. Most impressive to me is the phenomenal engineering that went into the theatre with a 2-million-gallon water tank for a stage. Each night the Cirque du Soleil performs an opera called "O," which involves dancing on the stage and then diving into the same spot that has miraculously opened up. The frosting on the cake is that "O" is taken from the French word *eau*, which means water.

In the middle of the Mohave Desert, the city is almost a shrine to water.

It is into this ostentatious display that Patricia Mulroy stepped in 1989. She was surrounded by water masters who blatantly showed off water with absolute confidence. The problems she faced were greater than the water displays. Ninety percent of Las Vegas' water comes from Lake Mead. By law only four percent of the lake's outtake can go to Las Vegas. The lion's share goes to Arizona and California.

The water level in Lake Mead is dropping dramatically, and the population of Las Vegas has been going up just as dramatically. Each of those new residents expects running water in his or her house. In addition, the city gets about 36 million visitors a year, each of whom expects all the conveniences water provides.

Adding insult to injury, amid the strains of growth and tourists, the rain and snowfall for the Colorado River, which feeds Lake Mead, has been dropping—again, dramatically. In

1990, the lake was 125 feet below what it had been a decade earlier. When you consider the lake is 110 miles long, that is a lot of lost water.

Into this water emergency, which Mulroy described as "organized chaos," she brought good water management and strong leadership. As a measure of her accomplishments, in 1989 when she became general manager of the Las Vegas Valley Water Authority, water consumption was 348 gallons per resident. Twenty years later, the average use per person was 240 gallons. Under her leadership, water consumption had dropped 31 percent.

Pricing was used as an incentive to reduce water use. She lowered the base rate, but instituted escalating rates based on quantity consumed.

Mulroy also went shopping for any ground water in the region that was unclaimed (a surprising 865,000 acre feet was found). Naturally, local people contested this "water theft" and battles in the court ensued.

Her aggressive tactics helped unify the seven Las Vegas area water utilities, and the Southern Nevada Water Authority (SNWA) was formed. She got the developer, Steve Wynn, to double plumb Treasure Island so he could use recycled water. This set a precedent. It also laid the groundwork for stiff guidelines for those wanting lavish water displays: Use well water from your own property, use treated water, or offset the water feature consumption by savings elsewhere.

The SNWA has a long list of ways to reduce consumption, including reuse water for all golf courses, sophisticated xeriscaping, and the rather famous "cash for grass" program where owners are paid $1.00-1.50/sf for grass removal which is replaced with desert landscaping.

One industry in Las Vegas deserves a special callout because the numbers help prove the economical benefits of saving water. A large laundry operation bought and installed an $800,000 reuse system. It saves $2,000 *each day* on the water bill. Additional savings come from reduced energy and sewage

Figure 12-10. Planting Grass

charges. The payback was just over a year. The company wants to do more in other locations, but the recession has limited its capital to do so. Sounds like all the ingredients for a performance contract.

In testimony before the US Senate in 2009, Mulroy said, "We know that the way we've been managing water resources for the last hundred years is obsolete." She points to the possessive, protectionist attitudes we have about water and pipelines, and questions why we can have transcontinental oil and gas pipelines, but transcontinental water pipelines seem unthinkable.

If we shift gears to the global stage, the situation in Italy needs a careful look. The country has serious water problems that have been recognized, but the solution is at a standstill. The following case study was written by James D. Hansen, a well-recognized writer and businessman in Italy. He is the owner of the consulting firm, HansenWorldwide, in Milan, Italy.

CASE STUDY: ITALY

In many parts of the world, public ownership and operation of water supplies is the most common circumstance. There are many reasons for this, some good and some bad. On one hand, the very high levels of capital investment required to build and maintain water supply infrastructure make government involvement a practical necessity. On the other, placing the water-supply network in political hands opens its management to many aspects of "good enough for government work" inefficiency.

Recognition of the problem in Italy led to a wave of utility privatizations in the two decades preceding the severe worldwide economic recession conventionally dated as beginning with the collapse of the bank Lehman Brothers in 2008. Economic dislocation and the fear that perhaps capitalism itself was somehow failing led to attempts to roll back the process of privatization in many Western countries.

The water utilities were particularly vulnerable in this respect because the availability of a high-quality water supply is now often considered a "right" of all citizens—possibly on the model of air-to-breathe being a human right, though air of course is self-delivering.

A case in point illustrating most of these phenomena is the Italian referendum of 2011, intended to return the supply of water to public hands. Though most of the system never left public ownership, significant pieces of it had been sold off during the Italian government's desperate efforts to clean up its balance sheets preparatory to entry into the "Euro" currency system in January 2002.

Though most Italians today have access to good-quality drinking water, the national system is notoriously inefficient. According to official data, as much as a third of all the water transiting the system is lost to leakage and other waste, and loss is estimated to reach points as high as 50 percent in some parts of southern Italy. These are the worst values in the European Union.

Privatization, however, had the effect of raising the cost of residential water supplies in those areas where it had been introduced, in large part because the political pricing of preceding decades simply could not be maintained in the face of the investments required to revamp infrastructure and to remunerate private investment.

The issue became a political football, and on the 12th and 13th of June 2011, Italians went to the polls to vote on a referendum principally intended to abrogate the legislation that had allowed the privatization of parts of the national water system.

Not quite 55% of the voting population participated in the referendum, and the result was overwhelmingly in favor of abrogation—a return to the publicly owned past. Approval rates all across the country were well above 90%, something more than a landslide.

This, of course, created a huge problem in concretely managing the water supply because the cards on the table had undergone a sudden and dramatic overnight change. The "official reaction"—if it could be called that—was typically Italian. First, a vast number of lawsuits and inconclusive public consultations tied the reform in knots, blocking the process.

Then, changes to administrative terminology were introduced to address the "letter of the law" but not its substance. For instance, the mechanism governing the generation of profit within the system was recast to describe the margin generated as "financial costs" rather than the "remuneration of operating capital"—a move intended to satisfy the public perception that "no-one should be allowed to profit" from the management of a public good like water. The actual level of remuneration (7%) was not changed.

In this way the political will was satisfied, though perhaps not so much that of the public. Further—and essentially—capital investment has been paralyzed. Public administration does not have the money to invest, and private sources of capital no longer wish to touch the water utility sector with a 10-foot pole, given the uncertainly surrounding it.

An important point in the Italian case study was the raised rates to fix the system and the public's outcry against it. Part of the problem, of course, was (and is) the economic conditions in Italy. The country, however, is now back to relying on the system that was not working all that well, and no relief is in sight.

ATLANTA: NOT A CASE STUDY

Coming back to the US, city planners might want to look at Atlanta as an example of *what not to do*. Just as the government controls Lake Mead and the allocations to Las Vegas are set by law, the government controls Lake Lanier, which supplies Atlanta's water. The government is committed to sustaining many uses from Lake Lanier, from a nuclear power plant in Alabama to the purple bankclimber in the Florida Panhandle. When conditions become even worse, who do you think will win? The Atlanta resident or the purple bankclimber? By the way, that's a mussel.

Unfortunately, most of the actions taken by Atlanta officials, dating back to Hartsfield, who declared Atlanta had plenty of water and would not participate in the Lake Lanier project, are examples of what not to do in city water management planning. It started with not buying a stake in the Lake Lanier effort and went downhill from there.

Atlanta is one of the fastest-growing cities in the country. From 2000 to 2009, it averaged 400 new residents a day. Each one of them needs, *and expects,* water.

FINAL THOUGHTS

Water poverty at any level is painful and ugly.

Water poverty is avoidable.

The Golden Age of Water is over; we have but to admit it and get to work.

After many years of research, Charles Fishman writes in *The Big Thirst,*

> You can't save a hundred gallons of water per day with posters and TV commercials urging people to take shorter showers and skip a car wash. You save that much water by thinking <u>twenty years ahead</u> and imagining what it would take to change how a whole community operates—how it thinks about itself—and by giving people time to think differently about themselves, their community and their water. [underscore supplied] [12-1]

It's interesting that Fishman sites 20 years as the time it takes. Estimates as to when the Ogallala aquifer will dry up are *about 20 years.* The clock is ticking.

Have we really considered the cost of running out of water or even having it in short supply? What happens to the economy? To the very way we live? How many jobs would be lost in your community with a 50 percent drop in water supply? Will the children get a bath? Do we really want to find out what water poverty looks like?

The sudden rush to take measures in California is eerily and hauntingly familiar. It's today's equivalent of the boarding up of school windows and blocking air intakes. Once more a vital resource is threatened and few, especially our public agencies, seem to know what to do.

Our abundance of cheap, safe water is disappearing. We are going to have to learn to reduce our demands and compromise our standards. We can desalinate, but it won't be cheap. We can reuse and install double plumbing and bring in the needed safeguards. We can learn to categorize water by fit to purpose and price accordingly while we figure a way to use each category more efficiently.

When water is finally regarded as the precious commodity it is, we will be more conscious of our use and consume the good stuff more sparingly. Even little things like turning off the water while we brush our teeth and the routine use of the "military" shower will become a way of life. We can actually learn from the villages where girls walk more than a mile to get water each day. When it is that hard to get, each drop is examined for multiple applications.

One of the joys of water is that it never goes away. We are still using the same water that was put on Earth 4 billion years ago. Unlike fossil fuels (which once used are gone) water is recycled for use again. And again. And again.

The problem is managing it. Tonight we water the garden. Much of that water evaporates and goes up to form clouds. While someone might need that water desperately, the clouds can just as easily drift away to someplace with plenty of water. We cannot control all that happens. We can be grateful, however, that Mother Nature does such a great job of cleaning it and sending it back so we can have another go at it.

As we continue to search for the best way to manage our water crisis, it is exceedingly frustrating to know that saving water in Seattle will not help a thirsty child in a small village in Africa. As we become more conscious of our water problems, we may become more sensitive to the people living in water poverty. Then, steps can be taken to lay a few pipes and make their water safe for drinking.

If you have reached this last page, it has become exceedingly obvious to you that while water problems are local and can best be addressed with local solutions, there is great danger in putting off much needed work. At the same time, you see that local water is part of a huge cleansing system. That's what makes it such a miracle. It's time we started treating it like one.

Appendices

APPENDIX A

WATER MEASUREMENT AND EQUIVALENCIES

1 gallon = 3.8 liters
1 gallon = 8.33 pounds
1 cubic meter – 264 gallons
1 cubic meter = 1,000 liters
1 cubic meter = 1 metric ton (2,200 pounds)
1 cubic foot = 7.48 gallons
1 cubic foot= 62 pounds
1 megaliter = 1,000,000 liters
1 megaliter = 264,000 gallons
1 gigaliter = 1,000,000,000 liters
1 gigaliter = 264,000,000 gallons
1 acre foot = 325,851 gallons
1 acre foot = 1.2 megaliters
1 acre foot = 7.48 gallons/sf
1 cubic kilometer = 15 trillion gallons

As of 2005, the average daily use per person in the US was 99 gallons and the average daily use per household was 262 gallons.

APPENDIX B

A PROCLAMATION OF A STATE OF EMERGENCY

WHEREAS the State of California is experiencing record dry conditions, with 2014 projected to become the driest year on record; and

WHEREAS the state's water supplies have dipped to alarming levels, indicated by: snowpack in California's mountains is approximately 20 percent of the normal average for this date; California's largest water reservoirs have very low water levels for this time of year; California's major river systems, including the Sacramento and San Joaquin rivers, have significantly reduced surface water flows; and groundwater levels throughout the state have dropped significantly; and

WHEREAS dry conditions and lack of precipitation present urgent problems: drinking water supplies are at risk in many California communities; fewer crops can be cultivated and farmers' long-term investments are put at risk; low-income communities heavily dependent on agricultural employment will suffer heightened unemployment and economic hardship; animals and plants that rely on California's rivers, including many species in danger of extinction, will be threatened; and the risk of wildfires across the state is greatly increased; and

WHEREAS extremely dry conditions have persisted since 2012 and may continue beyond this year and more regularly into the future, based on scientific projections regarding the impact of climate change on California's snowpack; and

WHEREAS the magnitude of the severe drought conditions presents threats beyond the control of the services, personnel, equipment and facilities of any single local government and require the combined forces of a mutual aid region or regions to combat; and

WHEREAS under the provisions of section 8558(b) of the California Government Code, I find that conditions of extreme peril to the safety of persons and property exist in California due to water shortage and drought conditions with which local authority is unable to cope.

NOW, THEREFORE, I, EDMUND G. BROWN JR., Governor of the State of California, in accordance with the authority vested in me by the state Constitution and statutes, including the California Emergency Services Act, and in particular, section 8625 of the California Government Code **HEREBY PROCLAIM A STATE OF EMERGENCY** to exist in the State of California due to current drought conditions.

IT IS HEREBY ORDERED THAT:

1. State agencies, led by the Department of Water Resources, will execute a statewide water conservation campaign to make all Californians aware of the drought and encourage personal actions to reduce water usage. This campaign will be built on the existing Save Our Water campaign (www. saveourh2o.org) and will coordinate with local water agencies. This campaign will call on Californians to reduce their water usage by 20 percent.

2. Local urban water suppliers and municipalities are called upon to implement their local water shortage contingency plans immediately in order to avoid or forestall outright restrictions that could become necessary later in the drought season. Local water agencies should also update their legally required urban and agricultural water management plans, which help plan for extended drought conditions. The Department of Water Resources will make

the status of these updates publicly available.

3. State agencies, led by the Department of General Services, will immediately implement water use reduction plans for all state facilities. These plans will include immediate water conservation actions, and a moratorium will be placed on new, non-essential landscaping projects at state facilities and on state highways and roads.

4. The Department of Water Resources and the State Water Resources Control Board (Water Board) will expedite the processing of water transfers, as called for in Executive Order B-21-13. Voluntary water transfers from one water right holder to another enables water to flow where it is needed most.

5. The Water Board will immediately consider petitions requesting consolidation of the places of use of the State Water Project and Federal Central Valley Project, which would streamline water transfers and exchanges between water users within the areas of these two major water projects.

6. The Department of Water Resources and the Water Board will accelerate funding for water supply enhancement projects that can break ground this year and will explore if any existing unspent funds can be repurposed to enable near-term water conservation projects.

7. The Water Board will put water right holders throughout the state on notice that they may be directed to cease or reduce water diversions based on water shortages.

8. The Water Board will consider modifying requirements for reservoir releases or diversion limitations, where existing requirements were established to implement a water qual-

ity control plan. These changes would enable water to be conserved upstream later in the year to protect cold water pools for salmon and steelhead, maintain water supply, and improve water quality.

9. The Department of Water Resources and the Water Board will take actions necessary to make water immediately available, and, for purposes of carrying out directives 5 and 8, Water Code section 13247 and Division 13 (commencing with section 21000) of the Public Resources Code and regulations adopted pursuant to that Division are suspended on the basis that strict compliance with them will prevent, hinder, or delay the mitigation of the effects of the emergency. Department of Water Resources and the Water Board shall maintain on their websites a list of the activities or approvals for which these provisions are suspended.

10. The state's Drinking Water Program will work with local agencies to identify communities that may run out of drinking water, and will provide technical and financial assistance to help these communities address drinking water shortages. It will also identify emergency interconnections that exist among the state's public water systems that can help these threatened communities.

11. The Department of Water Resources will evaluate changing groundwater levels, land subsidence, and agricultural land fallowing as the drought persists and will provide a public update by April 30 that identifies groundwater basins with water shortages and details gaps in groundwater monitoring.

12. The Department of Water Resources will work with counties to help ensure that well drillers submit required groundwater well logs for newly constructed and deepened wells

in a timely manner and the Office of Emergency Services will work with local authorities to enable early notice of areas experiencing problems with residential groundwater sources.

13. The California Department of Food and Agriculture will launch a one-stop website (www.cdfa.ca.gov/drought) that provides timely updates on the drought and connects farmers to state and federal programs that they can access during the drought.

14. The Department of Fish and Wildlife will evaluate and manage the changing impacts of drought on threatened and endangered species and species of special concern, and develop contingency plans for state Wildlife Areas and Ecological Reserves to manage reduced water resources in the public interest.

15. The Department of Fish and Wildlife will work with the Fish and Game Commission, using the best available science, to determine whether restricting fishing in certain areas will become necessary and prudent as drought conditions persist.

16. The Department of Water Resources will take necessary actions to protect water quality and water supply in the Delta, including installation of temporary barriers or temporary water supply connections as needed, and will coordinate with the Department of Fish and Wildlife to minimize impacts to affected aquatic species.

17. The Department of Water Resources will refine its seasonal climate forecasting and drought prediction by advancing new methodologies piloted in 2013.

18. The California Department of Forestry and Fire Protection

will hire additional seasonal firefighters to suppress wild-fires and take other needed actions to protect public safety during this time of elevated fire risk.

19. The state's Drought Task Force will immediately develop a plan that can be executed as needed to provide emergen-cy food supplies, financial assistance, and unemployment services in communities that suffer high levels of unem-ployment from the drought.

20. The Drought Task Force will monitor drought impacts on a daily basis and will advise me of subsequent actions that should be taken if drought conditions worsen.

I FURTHER DIRECT that as soon as hereafter possible, this Proclamation be filed in the Office of the Secretary of State and that widespread publicity and notice be given of this Procla-mation.

IN WITNESS WHEREOF I have hereunto set my hand and caused the Great Seal of the State of California to be affixed this 17th day of January, 2014.

EDMUND G. BROWN JR.,
Governor of California

References

SPECIFIC REFERENCES BY CHAPTER

Chapter 4
4-1 Grossman, Lev. "A Star Is Born." *Time*, November 2, 2015

Chapter 5
5-1 Burroughs, H.E. and Shirley J. Hansen, *Managing Indoor Air Quality. Fifth Edition*. Published by The Fairmont Press, Lilburn, Georgia. 2011

Chapter 6
6-1 United Nations, "Sustainable Development Goals: 17 Goals to Transform our World," 2015.
6-2 Hansen, Shirley J., *Manual for Intelligent Energy Services*. Published by The Fairmont Press, 2002
6-3 Hansen, Shirley J. and James W. Brown, *Sustainability Management Handbook*. Published by The Fairmont Press, 2011.
6-4 ConAgra, *ConAgra Food Sustainability Vision*, May 6, 2014

Chapter 7
7-1 McCardell, Sandra, "Energy & Systems/Systems & Energy—Introduction." *Current C Energy Systems, Inc, Newsletter*. April 30, 2015
7-2 Woodruff, Eric and Al Thumann, *How to Finance Energy Management Projects*. Published by The Fairmont Press, 2010
7-3 Red Mountain Insights, *Venture Capital for Energy Company Directory*, 2015. ISBN 978-1-62484-038-8
7-4 Langlois, Pierre and Shirley J. Hansen, *World ESCO Outlook*, Published by The Fairmont Press, Lilburn, Georgia. 2012
7-5 Hansen, Shirley J. and Jeannie Weisman, *Performance Contracting: Expanding Horizons*. Published by The Fairmont Press, Lilburn, Georgia. 1998

Chapter 8
8-1 Hofmeister, John, *Why We Hate the Oil Companies*. Published by Palgrave, Macmillan. 2010

8-2   Urdan, Mathew S., "Water Wars or Water Peace? Part 1" Analysis, *Foreign Relations* February 8, 2011

Chapter 10
10-1 Shiklomanov, Igor, *Water in Crisis.* Edited in 1993 by Peter Gleick
10-2 Smith, Shepherd, "Shepherd Smith Reporting" Fox News Channel, March 2016
10-3 Auvermann, Brent, Ogalalla Aquifer project, AgriLife, Texas A&M. March, 2016
10-4 Hansen, Shirley J. and James W. Brown, *Sustainability Management Handbook, op. cit.*
10-5 Troesken, Werner, *The Great Lead Water Pipe Disaster.* MIT Press, 2008

Chapter 11
11-1 Young, Rachel and Eric Mackres, American Council for an Energy-Efficient Economy and the Alliance for Water Efficiency. "Tackling the Nexus: Exemplary Programs that Save Energy and Water." Research Report E131, 2013
11-2 Young, Alison and Mark Nichols, " High Lead Levels Found in 2,000 Water Systems Across USA," *USA Today.* March 17, 2016
11-3 Hansen, Shirley J., "No Water Is as Costly as No Water," *Strategic Planning for Energy and the Environment.* Association of Energy Engineers, Atlanta Georgia. 2012

Chapter 12
12-1 Dow's Water & Process Solutions Division, The Dow Chemical Company, Midland, Michigan 2015
12-2 *International Performance Measurement and Verification Protocol,* Efficiency Valuation Organization, Washington, D.C. evoworld.org
12-3 Grady, Barbara, "Dow, Dais Analytic and Water Innovation," *GreenBiz Group*, October 26, 2015
12-4 Fishman, Charles, *The Big Thrist: The Secret Life and Turbulant Future of Water.* Free Press, 2012

GENERAL REFERENCES

Burroughs, H.E. and Shirley J. Hansen, *Managing Indoor Air Quality, Fifth Edition*. The Fairmont Press, 2011

CDC. *Violations identified from routine swimming pool inspections—selected states and counties, United States*. MMWR Morb Mortal Wkly Rep. 2010;59(19):582-7. 2008

CDC. *Surveillance data from public spa inspections—United States, May–September*. MMWR Morb Mortal Wkly Rep. 2004;53(25):553-5 2002

Curley, Emily "The End of an Era: Say Goodbye to Cheap Water," AtSite

Fishman, Charles, *The Big Thirst*. Free Press, 2012

Hansen, Shirley J. and James W. Brown. *Sustainability Management Handbook*. The Fairmont Press, 2011

Hansen, Shirley J. and Pierre Langlois, *World ESCO Outlook*. The Fairmont Press, 2012

Hlavsa MC, Roberts VA, Kahler AM, Hilborn ED, Wade TJ, Backer LC, Yoder, JS. *Recreational water–associated disease outbreaks—United States, 2009–2010*. MMWR Morb Mortal Wkly Rep. 2014;63(1):6-10.

Hydroponics (no author) "Myth: Hydroponics is bad for the environment" web 7 December 2015

Leonard, Tom. "Fog Harvesting and Dew Harvesting." Aug. 2009. http://fogharvesting.com

Pearce, Fred, *When the Rivers Run Dry*. Beacon Press, 2007.

Reisner, Marc, *Cadillac Desert: The American West and Its Disappearing Water*. Penguin books, 1993

Troesken, Werener, *The Great Lead Water Pipe Disaster*. MIT Press, 2008

Sedlak, David, *Water 4.0*, Yale University Press, 2015.

Siegel, Seth, *Let There Be Water: Israel's Solution for a Water-Starved World*. Thomas Dunne Books, Martin's Press, LLC 2015

Shields JM, Hill VR, Arrowood MJ, Beach MJ. Inactivation of Cryptosporidium Parvum under Chlorinated Recreational Water Conditions. *Water Health* 2008;6(4):513–20.

Slaughter, Carolyn and Paul Zummo "Coal: Not Going Away Anytime Soon." *Energybiz*, Volume 12//Issue 3, Summer 2015.

Solomon, Steven, *Water: The Epic Struggle for Wealth, Power and Civilization*. Harper Collins, 2011

Wise, Lindsay. "Ogallala Aquifer is Being Pumped dry: Great Plains water crisis is invisible, but deep." McClatchy, *Tacoma News Tribune*, August 2, 2015

Index